Vegetable, Starch & Pasta Station

NATIONAL APPRENTICESHIP TRAINING PROGRAM FOR COOKS

AMERICAN TECHNICAL PUBLISHERS
ORLAND PARK, ILLINOIS 60467-5756

American Culinary Federation

The *National Apprenticeship Training Program for Cooks Vegetable, Starch & Pasta Station* module contains procedures commonly practiced in the foodservice industry. Specific procedures vary with each task and must be performed by a qualified person. For maximum safety, always refer to specific manufacturer recommendations, insurance regulations, specific job site procedures, applicable federal, state, and local regulations, and any authority having jurisdiction. The material contained is intended to be an educational resource for the user. American Technical Publishers, Inc. assumes no responsibility or liability in connection with this material or its use by any individual or organization.

Cover Photo: Vulcan-Hart, a division of the ITW Food Equipment Group LLC

American Technical Publishers, Inc., Editorial Staff

Editor in Chief:
    Jonathan F. Gosse
Vice President—Production:
    Peter A. Zurlis
Director of Product Development:
    Cathy A. Scruggs
Art Manager:
    James M. Clarke
Multimedia Manager:
    Carl R. Hansen
Technical Editors:
    Cathy A. Scruggs
    Larry E. Pierce
Copy Editor:
    Catherine A. Mini

Cover Design:
    Jennifer M. Hines
Illustration/Layout:
    Jennifer M. Hines
    Melanie G. Doornbos
    Samuel T. Tucker
CD-ROM Development:
    Gretje Dahl
    Nicole S. Polak
    Daniel Kundrat

Printed in the United States of America

    ISBN 978-0-8269-4188-6

 This book is printed on recycled paper.

Acknowledgments

The American Culinary Federation Education Foundation (ACFEF) apprenticeship program and all its success are the result of the efforts of many individuals who work for the betterment of our apprentices and our organization as a whole. First and foremost, we thank the apprentices. Without their dedication and passion for becoming the next generation of professional chefs, we would have no goal to strive for. They work long hours in tough environments to become better culinarians. Their determination does not go unnoticed and is often what fuels the rest of us.

Just as importantly, we acknowledge the steadfast commitment of our many supervising chefs and program coordinators. They dedicate countless hours of time and effort to enriching the education of numerous apprentices, often without additional compensation or recognition. They tirelessly lead a new generation of culinarians to professional enlightenment.

No acknowledgment would be complete without mention of the many volunteers who make up the ACFEF National Apprenticeship Committee. These volunteers were invaluable in crafting and honing the current ACFEF apprenticeship program. Their time commitment was exceeded only by the value of their combined expertise. Without them, none of this would be possible.

American Technical Publishers and the American Culinary Federation sincerely appreciate the content expert who assisted with the development of this module.

Michael J. McGreal, CEC, CCE, CHE, FMP, CHA, MCFE
Department Chair, Culinary Arts & Hospitality Management
Joliet Junior College

American Technical Publishers and the American Culinary Federation also appreciate the experts who reviewed this module.

David W. Weir, MBA, CEC, CCE
Assistant Professor/Chef Instructor
Daytona State College
School of College of Hospitality and Culinary Management

Michael W. Riggs, PhD, CEC, CCA, FMP
Associate Professor of Culinary Arts
Bowling Green Technical College

American Technical Publishers and the American Culinary Federation are grateful for the technical assistance and images provided by the following companies, organizations, and individuals:

American Metalcraft, Inc.
Barilla America, Inc.
Basic American Foods
Browne-Halco (NJ)
California Strawberry Commission
Carlisle FoodService Products
Canadian Beef, Beef Information Centre
Daniel NYC
Dexter-Russell, Inc.
Eloma Combi Ovens
Florida Department of Agriculture and Consumer Services
Florida Department of Citrus
Florida Tomato Committee
Frieda's Specialty Produce
House Foods

Idaho Potato Commission
Indian Harvest Specialtifoods, Inc./Rob Yuretich
InterMetro Industries Corporation
Irinox USA
Kyocera Advanced Ceramics
McCain Foods USA
Melissa's Produce
Mercer Cutlery
Merrychef
Messermeister
Mushroom Council
National Cancer Institute, Daniel Sone (photograph)
National Cherry Growers and Industries Foundation
National Chicken Council

National Garden Bureau Inc.
National Honey Board
The National Pork Board
National Watermelon Promotion Board
Oregon Raspberry & Blackberry Commission
Paderno World Cuisine
Pear Bureau Northwest
Perdue Foodservice, Perdue Farms Incorporated
Tanimura & Antle®
United States Department of Agriculture
United States Potato Board
U.S. Apple Association
USA Rice Federation

Contents

VEGETABLE, STARCH & PASTA STATION
NATIONAL APPRENTICESHIP TRAINING PROGRAM FOR COOKS

Contents

INTERACTIVE CD-ROM CONTENTS

- Quick Quiz®
- Illustrated Glossary
- Flash Cards
- Media Clips
- Checkpoints

- Culinary Math Applications
- Certification Exam Preparation
- Apprenticeship Online Portal
- ATPeResources.com

Introduction

Welcome to the American Culinary Federation Education Foundation (ACFEF) apprenticeship program. Whether you are registered as an ACFEF apprentice or have a thirst for learning more, the culinary techniques outlined in this book have been validated by the culinary industry and sanctioned by the U.S. Department of Labor and the ACFEF National Apprenticeship Committee.

The ACFEF apprenticeship program focuses on hands-on training that allows apprentices to learn while being mentored by leading chefs in the industry. The program is rigorous and requires apprentices to complete 445 hours of classroom instruction with a minimum of 4,000 working hours in the kitchen. From the first to last day of the apprenticeship program, ACFEF apprentices are fully immersed in the smells, tastes, and textures of their culinary creations.

Several different options for the apprenticeship program are offered to meet the specific needs of apprentices. Program options range from two to three years and include the required classroom hours. Training is organized into 10 different stations, allowing apprentices to work at their own pace. It is an "earn while you learn" approach to training. Graduates from the ACFEF apprenticeship program earn recognition as an apprentice from the U.S. Department of Labor and the American Culinary Federation (ACF) and are given the opportunity to test for industry-recognized certification.

The ACFEF apprenticeship program also addresses sustainability. The quality of the food a chef serves is directly related to the health of the ecosystem. The program encompasses a multitude of topics from recycling and composting to purchasing locally and organically.

The journey to becoming a chef can be challenging and at times may seem difficult. But there is no substitute for valuable experience. Best of luck to you as you embark on your way to becoming a chef.

The ACFEF Apprenticeship Team

THE VEGETABLE, STARCH & PASTA STATION

The third station in the ACFEF apprenticeship program encompasses a wide variety of techniques related to the preparation of vegetable, starch, and pasta products. A major learning component of this station is the identification of various types of vegetables, sea vegetables, mushrooms, fruits, potatoes, grains, and pastas. Another important learning component of this station is knife skills.

Upon completion of this station, you will be able to identify products as the edible root, bulb, tuber, stem, leaf, flower, or seed of a plant. You will also be able to identify different types of starches and varieties of pasta. Another critical component of this station is knife safety and developing a proficient use of knives and specialty cutting tools.

Once you are well-versed and knowledgeable about each type of food product and basic cutting techniques, you will learn how to use appropriate cooking techniques to prepare vegetables, fruits, starches, and pastas in the professional kitchen and how to properly store these products. You will also learn how to plate these foods to help create balance and nutrition.

Finally, you will learn many sustainable practices at this station that will put you one step closer to completing the ACFEF apprenticeship program. Sustainable practices include using a power-up/power-down schedule for high-energy equipment and replacing incandescent light bulbs with fluorescent bulbs. Also, using ENERGY STAR™ rated equipment in the professional kitchen will conserve energy and result in substantial savings.

American Culinary Federation
Education Foundation

Features

MODULE FEATURES

The *National Apprenticeship Training Program for Cooks Vegetable, Starch & Pasta Station* module includes several features to make learning easier.

Procedures list the steps required to prepare a specific food. Complex procedures often include illustrations.

Certification Exam Preparation Questions, located at the end of the module, include 20 multiple-choice items, 4 essay items, and 1 sketching activity.

Sustainability Corner addresses content related to topics such as recycling, composting, energy conservation, and product repurposing. A QR code is included with this feature.

Media Clip icons indicate that an applicable media clip can be referenced on the Interactive CD-ROM.

Production Tips highlight ways to work efficiently in the professional kitchen.

QR codes link to Internet resources and are directly tied to the topic by which they appear.

Checkpoints appear throughout the module, and each one serves as a comprehension check of a specific portion of the module content.

Key Terms and their definitions are listed on the page prior to the "Sustainability Corner." Terms are also italicized at their point of definition within the module.

Nutrition Notes highlight key nutrition information about specific foods.

INTERACTIVE CD-ROM FEATURES

The *National Apprenticeship Training Program for Cooks Vegetable, Starch & Pasta Station Interactive CD-ROM* is a self-study aid that reinforces content. This CD-ROM is Windows® compatible.

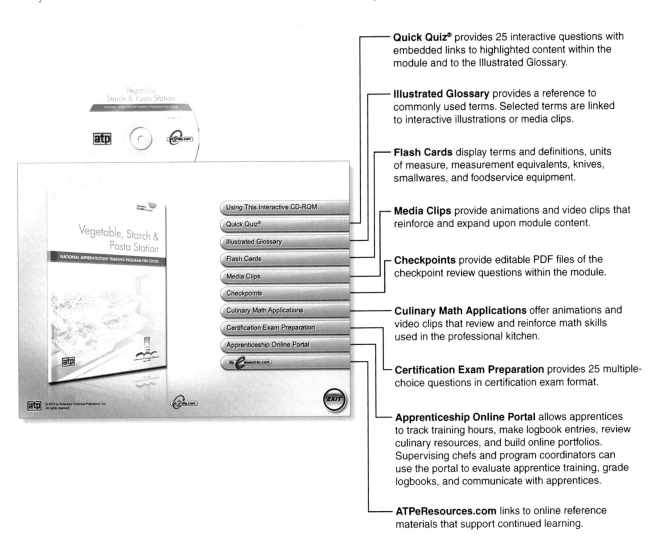

Quick Quiz® provides 25 interactive questions with embedded links to highlighted content within the module and to the Illustrated Glossary.

Illustrated Glossary provides a reference to commonly used terms. Selected terms are linked to interactive illustrations or media clips.

Flash Cards display terms and definitions, units of measure, measurement equivalents, knives, smallwares, and foodservice equipment.

Media Clips provide animations and video clips that reinforce and expand upon module content.

Checkpoints provide editable PDF files of the checkpoint review questions within the module.

Culinary Math Applications offer animations and video clips that review and reinforce math skills used in the professional kitchen.

Certification Exam Preparation provides 25 multiple-choice questions in certification exam format.

Apprenticeship Online Portal allows apprentices to track training hours, make logbook entries, review culinary resources, and build online portfolios. Supervising chefs and program coordinators can use the portal to evaluate apprentice training, grade logbooks, and communicate with apprentices.

ATPeResources.com links to online reference materials that support continued learning.

Vegetable, Starch & Pasta Station

Vegetables, fruits, potatoes, grains, and pastas are versatile foods, and it is important to be able to identify their many varieties. Knowledge of the market forms of these foods and how they are stored in the professional kitchen is also necessary. Equally important is the ability to correctly use large and small knives and cutting tools while preparing these foods. Many are prepped before the day of service and then finished using a variety of cooking methods. When knowledge of these foods and proficient knife skills have been acquired, an apprentice will be well prepared to tackle a multitude of recipes, from simple to complex.

VEGETABLE CLASSIFICATIONS

A *vegetable* is an edible root, bulb, tuber, stem, leaf, flower, or seed of a nonwoody plant. **See Figure 3-1.** Each vegetable offers culinary professionals unique opportunities to enhance menus with healthy, earthy flavors. Most vegetables can be served raw or cooked. Vegetables are an excellent source of vitamins, minerals, and fiber. They are also low in fat and calories. Vegetables have the best flavor and color when purchased in season.

Foodservice professionals should be able to identify common root, bulb, tuber, stem, leaf, flower, and seed vegetables. Likewise, foodservice professionals should be able to identify common fruit-vegetables, sea vegetables, and mushrooms that are prepared and served like vegetables.

Nutrition Note

Eating five servings per day of vegetables and fruits is optimal for weight management and chronic disease prevention. Providing a wide variety of vegetables on the menu can increase guest consumption of these nutritious foods.

Vegetables

Figure 3-1. A vegetable is an edible root, bulb, tuber, stem, leaf, flower, or seed of a nonwoody plant.

Root Vegetables

A *root vegetable* is an earthy-flavored vegetable that grows underground and has leaves that extend above ground. Root vegetables include beets, carrots, celeriac, jicamas, lotus roots, parsnips, radishes, rutabagas, salsify, and turnips. **See Figure 3-2.** *Note:* Ginger and horseradish are also edible roots, but they are not classified as vegetables.

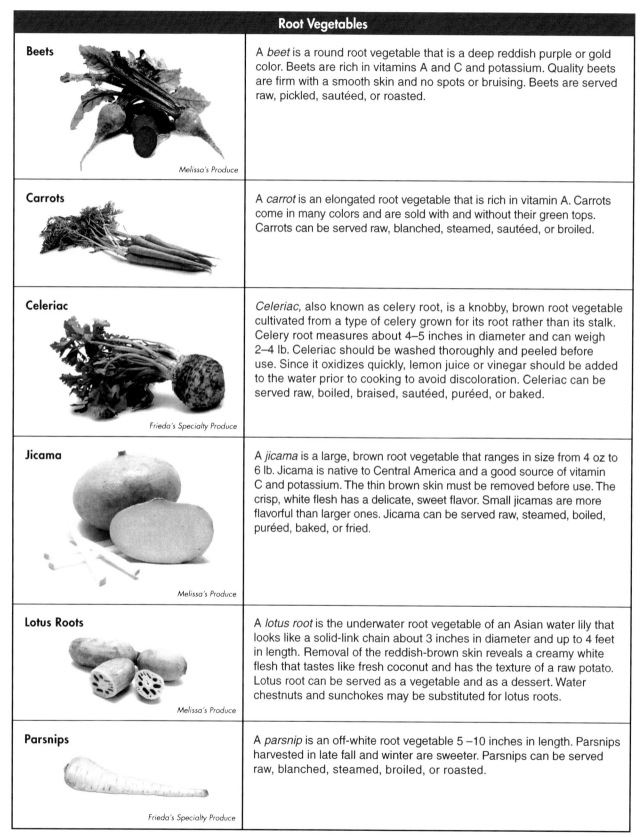

Root Vegetables	
Beets *Melissa's Produce*	A *beet* is a round root vegetable that is a deep reddish purple or gold color. Beets are rich in vitamins A and C and potassium. Quality beets are firm with a smooth skin and no spots or bruising. Beets are served raw, pickled, sautéed, or roasted.
Carrots	A *carrot* is an elongated root vegetable that is rich in vitamin A. Carrots come in many colors and are sold with and without their green tops. Carrots can be served raw, blanched, steamed, sautéed, or broiled.
Celeriac *Frieda's Specialty Produce*	*Celeriac,* also known as celery root, is a knobby, brown root vegetable cultivated from a type of celery grown for its root rather than its stalk. Celery root measures about 4–5 inches in diameter and can weigh 2–4 lb. Celeriac should be washed thoroughly and peeled before use. Since it oxidizes quickly, lemon juice or vinegar should be added to the water prior to cooking to avoid discoloration. Celeriac can be served raw, boiled, braised, sautéed, puréed, or baked.
Jicama *Melissa's Produce*	A *jicama* is a large, brown root vegetable that ranges in size from 4 oz to 6 lb. Jicama is native to Central America and a good source of vitamin C and potassium. The thin brown skin must be removed before use. The crisp, white flesh has a delicate, sweet flavor. Small jicamas are more flavorful than larger ones. Jicama can be served raw, steamed, boiled, puréed, baked, or fried.
Lotus Roots *Melissa's Produce*	A *lotus root* is the underwater root vegetable of an Asian water lily that looks like a solid-link chain about 3 inches in diameter and up to 4 feet in length. Removal of the reddish-brown skin reveals a creamy white flesh that tastes like fresh coconut and has the texture of a raw potato. Lotus root can be served as a vegetable and as a dessert. Water chestnuts and sunchokes may be substituted for lotus roots.
Parsnips *Frieda's Specialty Produce*	A *parsnip* is an off-white root vegetable 5 –10 inches in length. Parsnips harvested in late fall and winter are sweeter. Parsnips can be served raw, blanched, steamed, broiled, or roasted.

Figure 3-2. (continued on next page)

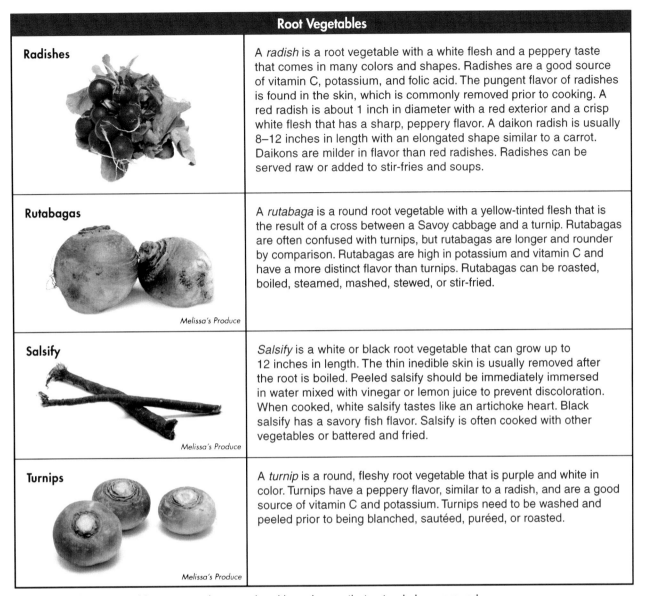

Root Vegetables	
Radishes	A *radish* is a root vegetable with a white flesh and a peppery taste that comes in many colors and shapes. Radishes are a good source of vitamin C, potassium, and folic acid. The pungent flavor of radishes is found in the skin, which is commonly removed prior to cooking. A red radish is about 1 inch in diameter with a red exterior and a crisp white flesh that has a sharp, peppery flavor. A daikon radish is usually 8–12 inches in length with an elongated shape similar to a carrot. Daikons are milder in flavor than red radishes. Radishes can be served raw or added to stir-fries and soups.
Rutabagas *Melissa's Produce*	A *rutabaga* is a round root vegetable with a yellow-tinted flesh that is the result of a cross between a Savoy cabbage and a turnip. Rutabagas are often confused with turnips, but rutabagas are longer and rounder by comparison. Rutabagas are high in potassium and vitamin C and have a more distinct flavor than turnips. Rutabagas can be roasted, boiled, steamed, mashed, stewed, or stir-fried.
Salsify *Melissa's Produce*	*Salsify* is a white or black root vegetable that can grow up to 12 inches in length. The thin inedible skin is usually removed after the root is boiled. Peeled salsify should be immediately immersed in water mixed with vinegar or lemon juice to prevent discoloration. When cooked, white salsify tastes like an artichoke heart. Black salsify has a savory fish flavor. Salsify is often cooked with other vegetables or battered and fried.
Turnips *Melissa's Produce*	A *turnip* is a round, fleshy root vegetable that is purple and white in color. Turnips have a peppery flavor, similar to a radish, and are a good source of vitamin C and potassium. Turnips need to be washed and peeled prior to being blanched, sautéed, puréed, or roasted.

Figure 3-2. Root vegetables grow underground and have leaves that extend above ground.

Nutrition Note

Radishes are high in fiber and water content and low in calories. Radish greens contain more vitamin C, calcium, and iron than the radish root.

Bulb Vegetables

A *bulb vegetable* is a strongly flavored vegetable that grows underground and consists of a short stem base with one or more buds that are enclosed in overlapping membranes or leaves. Bulb vegetables are very fragrant and are used for their aromatic qualities as well as their flavors. Examples of bulb vegetables include garlic, leeks, onions, ramps, scallions, and shallots. **See Figure 3-3.**

Tubers

A *tuber* is a short, fleshy vegetable that grows underground and bears buds capable of producing new plants. Examples of tubers include ocas, potatoes, sunchokes, sweet potatoes, water chestnuts, and yams. **See Figure 3-4.**

Bulb Vegetables	
Garlic *Eloma Combi Ovens*	*Garlic* is a bulb vegetable made up of several small cloves that are enclosed in a thin, husklike skin. White garlic is the most common variety and the most pungent in flavor. Pink garlic is named for its pinkish outer covering. Elephant garlic contains much larger cloves and is milder in flavor than smaller varieties. The flavor of garlic is released when a clove is cut, crushed, or minced and increases the more finely the clove is cut. Cut garlic must be refrigerated. Garlic is used to flavor a wide variety of dishes.
Leeks *Tanimura & Antle®*	A *leek* is a long, white bulb vegetable, with long, wide, flat leaves. Leeks are similar in appearance to scallions, but are much larger, milder, and sweeter than onions. The green leaves are often used to flavor soups and stocks, whereas the white portion of the leek is used in a variety of recipes.
Onions *Melissa's Produce*	An *onion* is a bulb vegetable made up of many concentric layers of fleshy leaves. Onion varieties include white, yellow, red, and pearl onions. The variety of the onion and the climate where it was grown determine how strong its flavor is. White onions have a slightly sweet flavor. Yellow onions, also known as Spanish onions, are very mild in flavor. Red onions are the sweetest variety and are commonly added to salads and sandwiches for color. Onions can be served raw, sautéed, grilled, roasted, stir-fried, or deep-fried.
Ramps *Melissa's Produce*	A *ramp* is a wild leek with a flavor similar to scallions, yet with more zing. Ramps are often diced for use in salads or on sandwiches. They can also be sautéed for use in egg or potato dishes.
Scallions *Tanimura & Antle®*	A *scallion,* also known as a green onion, is a small bulb vegetable with a slightly swollen base and long, slender, green leaves that are hollow. Scallions are mildly flavored compared to onions. Scallions are often added to salads or used as a garnish.
Shallots *Melissa's Produce*	A *shallot* is a very small bulb vegetable similar in shape to garlic and has two or three cloves inside. The outer covering can be bronze-colored, rose-colored, or pale gray. Shallots have a pink-tinged ivory flesh and a more subtle flavor than onions. When purchasing shallots, it is important to choose those that are firm and dry-skinned and to avoid any that are sprouting.

Figure 3-3. Bulb vegetables are used for their aromatic qualities as well as their flavors.

Tubers	
Ocas *Melissa's Produce*	An *oca,* also known as a New Zealand yam, is a small, knobby tuber that has a potato-like flesh and ranges in flavor from very sweet to slightly acidic. Ocas are white, pink, or red in color and are a good source of carbohydrates, calcium, dietary fiber, and iron. Ocas must be kept in a cool, dark place at room temperature or refrigerated in a crisper. Ocas can be served raw in salads, cooked like potatoes, or pickled.
Potatoes *Frieda's Specialty Produce*	A *potato* is a round, oval, or elongated tuber that is the only edible part of the potato plant. The color of potato skin differs among varieties and can be brown, red, yellow, white, orange, blue, or purple. Potato flesh can be creamy white to yellow-gold or purple in color. When purchasing potatoes, firm, undamaged potatoes with no signs of sprouting should be chosen. Potatoes must be stored in a dry, cool, dark place that allows them to breathe. If potatoes do not have adequate ventilation they quickly rot. Potatoes can be baked, sautéed, broiled, grilled, or fried.
Sunchokes *Frieda's Specialty Produce*	A *sunchoke,* also known as a Jerusalem artichoke, is a tuber with thin, brown, knobby-looking skin. The skin is edible but is often removed before cooking. The white flesh is crisp and sweet. Sunchokes can be blanched, steamed, puréed, or used to flavor soups.
Sweet Potatoes *Melissa's Produce*	A *sweet potato* is a tuber that has a paper-thin skin and flesh that ranges in color from ivory to dark orange. Sweet potatoes are an excellent source of vitamin A and potassium. The skin is edible, although it is often removed before cooking. Peeled or cut sweet potatoes oxidize, so it is important to place them in cold water until they are used. Sweet potatoes can be prepared in the same manner as potatoes. They also can be incorporated into breads and desserts.
Water Chestnuts *Frieda's Specialty Produce*	A *water chestnut,* also known as a water caltrop, is a small tuber with brownish-black skin and white flesh. A water chestnut is crunchy and juicy and resembles a chestnut in exterior color and shape. Water chestnuts have a mild, sweet flavor. Fresh water chestnuts must be peeled before use and may be refrigerated for up to one week if they are tightly wrapped. Canned water chestnuts are also available, but inferior in quality.
Yams *Melissa's Produce*	A *yam* is a large tuber that has thick, barklike skin and a flesh that varies in color from ivory to purple. The skin is inedible. Yams are commonly confused with sweet potatoes because they are often labeled as sweet potatoes in the United States, but they are different vegetables. Common varieties of yams include the tropical yam, garnet yam, and jewel yam. Yams are low in fat and a good source of carbohydrates, protein, and vitamins A and C. Yams can be prepared in the same manner as sweet potatoes.

Figure 3-4. Tubers grow underground and bear buds capable of producing new plants.

Stem Vegetables

A *stem vegetable* is the main trunk of a plant that develops buds and shoots instead of roots. Stem vegetables contain a lot of cellulose and become tougher as they continue to develop. Therefore, stems are usually harvested while tender. Examples of stem vegetables include asparagus, celery, fennel, hearts of palm, kohlrabi, and rhubarb. **See Figure 3-5.**

Stem Vegetables	
Asparagus	*Asparagus* is a green, white, or purple edible stem vegetable that is referred to as a spear. Green asparagus is the most common variety of asparagus. White asparagus is grown covered in soil to prevent photosynthesis from taking place and harvested as soon as the spears begin to emerge. White asparagus is more tender, but less flavorful, than green asparagus. Purple asparagus is sweeter than green asparagus because it has a higher sugar content. A cancer-fighting phytochemical called anthocyanin gives purple asparagus its purple hue. Raw asparagus is a popular ingredient in salads, omelets, quiches, and pasta dishes. Asparagus can be served raw, broiled, grilled, steamed, or puréed.
Celery *Tanimura & Antle®*	*Celery* is a green stem vegetable that has multiple stalks measuring 12–20 inches in length. The inner stalks, or hearts, are sweeter and more tender than the outer stalks. Celery should be shiny, firm, and crisp. Celery can be served raw, sautéed, stir-fried, or roasted.
Fennel *Melissa's Produce*	*Fennel* is a celery-like stem vegetable with overlapping leaves that grow out of a large bulb at its base. Fennel has a mild, sweet flavor that is often associated with licorice or anise. Fennel can be served raw, sautéed, broiled, blanched, or steamed.
Hearts of Palm *Melissa's Produce*	A *heart of palm* is a slender, white, stem vegetable that is surrounded by a tough husk. Hearts of palm are about 4 inches long and can be up to 1½ inches thick, although most are very thin. They are good sources of fiber and do not contain any cholesterol. Once the husks have been removed, hearts of palm can be served raw, steamed, or fried.
Kohlrabi *Frieda's Specialty Produce*	*Kohlrabi* is a sweet, crisp, stem vegetable that has a pale-green or purple, bulbous stem and dark-green leaves. Kohlrabi is created by crossbreeding a cabbage and a turnip. Although the entire kohlrabi is edible, the bulbous stem is the portion primarily used in cooking. The inner part of the stem base may be removed to produce a cavity that can be stuffed. Kohlrabi can be eaten raw, blanched, sautéed, or stir-fried.
Rhubarb *Frieda's Specialty Produce*	*Rhubarb* is a tart stem vegetable that ranges in color from pink to red and is usually prepared like a fruit. It may be peeled or left with the skin intact, depending on the use. Rhubarb leaves are never used because they contain a poisonous toxin called oxalate. Rhubarb can be sweetened and stewed to make sauces, but it is most commonly used to make pies, tarts, and other desserts.

Figure 3-5. Stem vegetables contain a lot of cellulose and become tougher as they continue to develop.

Leaf Vegetables

Leaf vegetables, also known as greens, are plant leaves that are often accompanied by edible leafstalks and shoots. Although edible leaves or greens can be eaten raw, they are often cooked to decrease their bitterness and increase their palatability. Leaf vegetables include beet greens, Belgian endive, bok choy, Brussels sprouts, butterhead lettuces, chard, chicory, collards, crisphead lettuces, dandelion greens, fiddlehead ferns, frisée, head cabbages, kale, looseleaf lettuces, mâche, mesclun greens, microgreens, mustard greens, Napa cabbages, nopales, radicchio, romaine lettuces, Savoy cabbages, sorrel, spinach, tatsoi, turnip greens, and watercress. **See Figure 3-6.**

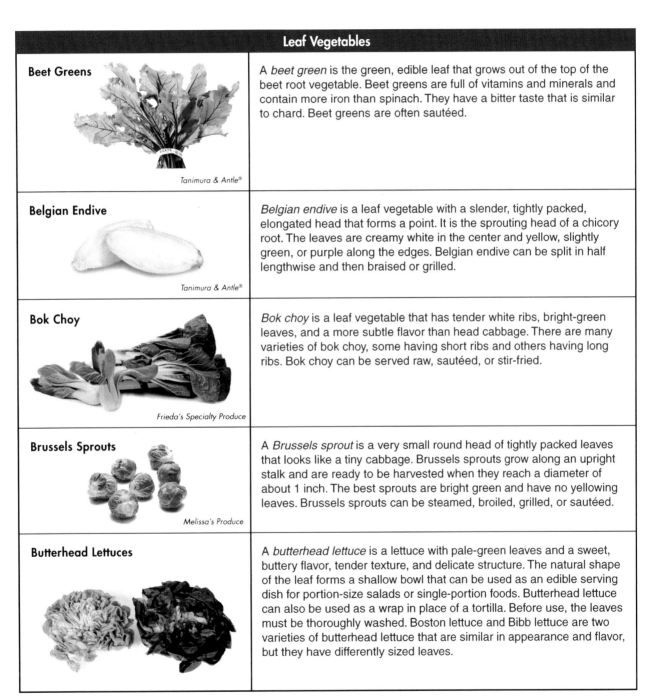

Leaf Vegetables	
Beet Greens *Tanimura & Antle®*	A *beet green* is the green, edible leaf that grows out of the top of the beet root vegetable. Beet greens are full of vitamins and minerals and contain more iron than spinach. They have a bitter taste that is similar to chard. Beet greens are often sautéed.
Belgian Endive *Tanimura & Antle®*	*Belgian endive* is a leaf vegetable with a slender, tightly packed, elongated head that forms a point. It is the sprouting head of a chicory root. The leaves are creamy white in the center and yellow, slightly green, or purple along the edges. Belgian endive can be split in half lengthwise and then braised or grilled.
Bok Choy *Frieda's Specialty Produce*	*Bok choy* is a leaf vegetable that has tender white ribs, bright-green leaves, and a more subtle flavor than head cabbage. There are many varieties of bok choy, some having short ribs and others having long ribs. Bok choy can be served raw, sautéed, or stir-fried.
Brussels Sprouts *Melissa's Produce*	A *Brussels sprout* is a very small round head of tightly packed leaves that looks like a tiny cabbage. Brussels sprouts grow along an upright stalk and are ready to be harvested when they reach a diameter of about 1 inch. The best sprouts are bright green and have no yellowing leaves. Brussels sprouts can be steamed, broiled, grilled, or sautéed.
Butterhead Lettuces	A *butterhead lettuce* is a lettuce with pale-green leaves and a sweet, buttery flavor, tender texture, and delicate structure. The natural shape of the leaf forms a shallow bowl that can be used as an edible serving dish for portion-size salads or single-portion foods. Butterhead lettuce can also be used as a wrap in place of a tortilla. Before use, the leaves must be thoroughly washed. Boston lettuce and Bibb lettuce are two varieties of butterhead lettuce that are similar in appearance and flavor, but they have differently sized leaves.

Figure 3-6. (continued on next page)

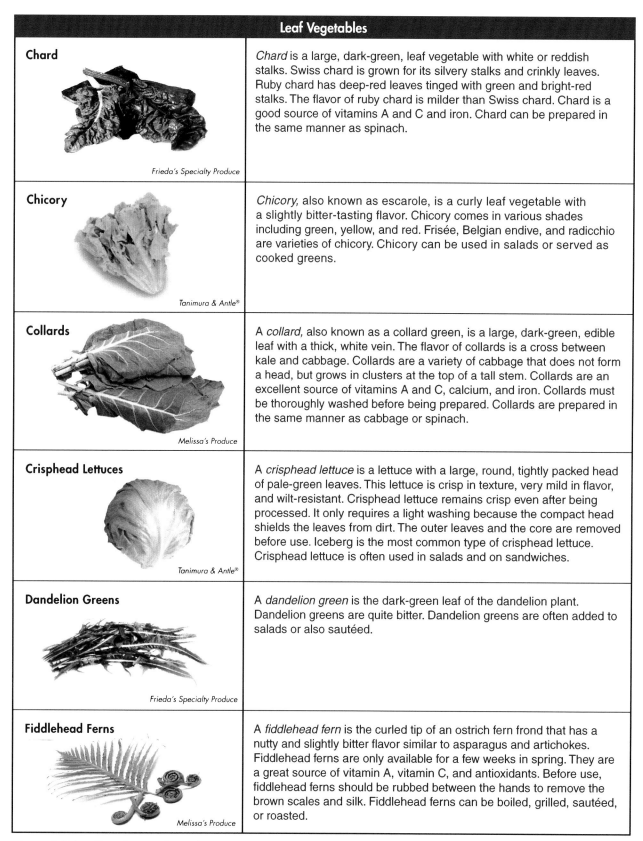

Leaf Vegetables	
Chard *Frieda's Specialty Produce*	*Chard* is a large, dark-green, leaf vegetable with white or reddish stalks. Swiss chard is grown for its silvery stalks and crinkly leaves. Ruby chard has deep-red leaves tinged with green and bright-red stalks. The flavor of ruby chard is milder than Swiss chard. Chard is a good source of vitamins A and C and iron. Chard can be prepared in the same manner as spinach.
Chicory *Tanimura & Antle®*	*Chicory,* also known as escarole, is a curly leaf vegetable with a slightly bitter-tasting flavor. Chicory comes in various shades including green, yellow, and red. Frisée, Belgian endive, and radicchio are varieties of chicory. Chicory can be used in salads or served as cooked greens.
Collards *Melissa's Produce*	A *collard,* also known as a collard green, is a large, dark-green, edible leaf with a thick, white vein. The flavor of collards is a cross between kale and cabbage. Collards are a variety of cabbage that does not form a head, but grows in clusters at the top of a tall stem. Collards are an excellent source of vitamins A and C, calcium, and iron. Collards must be thoroughly washed before being prepared. Collards are prepared in the same manner as cabbage or spinach.
Crisphead Lettuces *Tanimura & Antle®*	A *crisphead lettuce* is a lettuce with a large, round, tightly packed head of pale-green leaves. This lettuce is crisp in texture, very mild in flavor, and wilt-resistant. Crisphead lettuce remains crisp even after being processed. It only requires a light washing because the compact head shields the leaves from dirt. The outer leaves and the core are removed before use. Iceberg is the most common type of crisphead lettuce. Crisphead lettuce is often used in salads and on sandwiches.
Dandelion Greens *Frieda's Specialty Produce*	A *dandelion green* is the dark-green leaf of the dandelion plant. Dandelion greens are quite bitter. Dandelion greens are often added to salads or also sautéed.
Fiddlehead Ferns *Melissa's Produce*	A *fiddlehead fern* is the curled tip of an ostrich fern frond that has a nutty and slightly bitter flavor similar to asparagus and artichokes. Fiddlehead ferns are only available for a few weeks in spring. They are a great source of vitamin A, vitamin C, and antioxidants. Before use, fiddlehead ferns should be rubbed between the hands to remove the brown scales and silk. Fiddlehead ferns can be boiled, grilled, sautéed, or roasted.

Figure 3-6. (continued on next page)

Leaf Vegetables	
Frisée *Tanimura & Antle®*	*Frisée*, also known as curly endive, is a leaf vegetable with twisted, thin leaves that grow in a loose bunch. The leaves vary in color from dark green outer leaves to pale green or white inner leaves. Frisée has a bitter flavor and pairs well with strongly flavored cheeses and acidic vinaigrettes. It can also be used as a garnish.
Head Cabbages	*Head cabbage* is a tightly packed, round head of overlapping leaves that can be green, purple, red, or white in color. Head cabbage usually ranges from 2–8 pounds and from 4–10 inches in diameter. The base of the head where the leaves attach to the stalk is known as the heart. The inedible heart is removed during preparation. The best heads are heavy and compact. Head cabbage can be eaten raw, steamed, braised, roasted, or stir-fried.
Kale *Melissa's Produce*	*Kale* is a large, frilly, leaf vegetable that varies in color from green and white to shades of purple. Although all varieties of kale are edible, green varieties are better for cooking. Because of its bitterness, kale is rarely eaten raw. The center stalk is often removed before kale is cooked. Kale is an ample source of vitamins A and C, folic acid, calcium, and iron. Kale may be prepared in the same manner as spinach.
Looseleaf Lettuces *Tanimura & Antle®*	A *looseleaf lettuce* is a mild flavored, rich-colored lettuce with a cascade of leaves held loosely together at the root. Some varieties have thick leaves, and others have thin leaves. Some leaves are flat, while others are frilled or curled. Looseleaf lettuces are commonly used in salads and on sandwiches.
Mâche	*Mâche*, also called lamb's lettuce, is a small, leafy green with a velvety texture. Mâche can have dark-green, scoop-shaped leaves or elongated, pale-green leaves. It is tender and mild in flavor and pairs well with vinaigrettes.
Mesclun Greens	*Mesclun greens* are a mix of young greens that range in color, texture, and flavor. Mesclun greens can be tender and sweet or crisp and peppery. Mesclun is most often a mix of 10–12 varieties of greens, often including romaine, radicchio, endive, baby spinach, red oak, leaf lettuce, arugula, frisée, and tatsoi. Some mesclun mixes contain as many as 30 different plants, including flowers and herbs.

Figure 3-6. (continued on next page)

Leaf Vegetables	
Microgreens	*Microgreens* are the first sprouting leaves of an edible plant. In addition to being used in salads, microgreens are commonly used to garnish hors d'oeuvres. Common varieties of microgreens include beet greens, turnip greens, spinach, and kale.
Mustard Greens	A *mustard green* is a large, dark-green leaf vegetable from the mustard plant that has a strong peppery flavor. Mustard greens are an excellent source of vitamins A and C, thiamin, and riboflavin. Mustard greens must be thoroughly washed before being cooked. Mustard greens may be steamed, braised, sautéed, or stir-fried.
Napa Cabbages *Frieda's Specialty Produce*	*Napa cabbage,* also known as celery cabbage, is an elongated head of crinkly and overlapping edible leaves that are a pale yellow-green color with a white vein. Napa cabbage is often referred to as Chinese cabbage. Its leaves are more tender than those of head cabbage and it has a very delicate flavor. Napa cabbage is most often served raw or stir-fried.
Nopales *Melissa's Produce*	A *nopal* is the green, edible leaf of the prickly pear cactus. Nopales measure about 5 inches in length and 3–4 inches in width. They are crunchy, slippery, and slightly tart. Nopales are a great source of calcium and vitamins A and C. When preparing nopales the eyes, prickles, and all dry areas are removed. Nopales may be served raw, steamed, or sautéed.
Radicchio *Tanimura & Antle®*	*Radicchio* is a small, compact head of red leaves, similar to a small head of red cabbage. Radicchio leaves form a bowl shape and can hold individually sized salads. Radicchio is also commonly sautéed or braised and served as a side.
Romaine Lettuces *Tanimura & Antle®*	A *romaine lettuce* is a lettuce with long, green leaves that grow in a loosely packed, elongated head on crisp center ribs. The outer edges of the leaves are darker in color and lighten to a pale celadon near the rib. Romaine lettuce does not bruise easily when cut and is high in vitamin A, vitamin C, vitamin K, and folate. Dirt collects in the ridges of the loose leaves, so romaine lettuce must be washed thoroughly. Romaine lettuce has a mild, sweet flavor and blends well with other greens.

Figure 3-6. (continued on next page)

Leaf Vegetables	
Savoy Cabbages *Melissa's Produce*	*Savoy cabbage* is a conical-shaped head of tender, crinkly, edible leaves that are blue-green on the exterior and pale green on the interior. The leaves are very pliable and have a distinct sweet flavor. Savoy cabbages lack the sulfur-like odor often associated with cooking other cabbage varieties. Savoy cabbage can be stir-fried, stuffed, or used raw in salads.
Sorrel *Melissa's Produce*	*Sorrel* is a large, green, leaf vegetable that ranges in color from pale green to dark green and from 2–12 inches in length. Sorrel is quite acidic in flavor. Strongly flavored sorrel is called sour dock or sour grass. The acidic flavor comes from the presence of oxalic acid. Sorrel is a good source of vitamin A. Like spinach, sorrel can be served raw or cooked.
Spinach *Tanimura & Antle®*	*Spinach* is a dark-green leaf vegetable with a slightly bitter flavor that may have flat or curly leaves, depending on the variety. Fresh spinach is rich in vitamins A and C, folate, potassium, iron, and magnesium. Spinach is usually very gritty and must be thoroughly rinsed. The tough stems are usually removed before the spinach is cooked. Spinach can be served raw, steamed, or sautéed.
Tatsoi	*Tatsoi* is a spoon-shaped, emerald-colored leaf vegetable native to Japan. Tatsoi has a mild flavor and is a good source of vitamins, minerals, and antioxidants. Tatsoi may be served raw in salads, steamed, sautéed, or boiled.
Turnip Greens	A *turnip green* is a dark-green leaf vegetable that grows out of the top of the turnip root vegetable. Young turnip greens have a sweet flavor. As the plant ages, the leaves become bitter. Turnip greens are an excellent source of vitamins A and C and a good source of riboflavin, calcium, and iron. Turnip greens must be thoroughly washed before being cooked. The removal of the ribs from the leaves will yield a more tender batch of greens. Turnip greens may be steamed, braised, sautéed, or stir-fried.
Watercress *Tanimura & Antle®*	*Watercress* is a small, crisp, dark-green, leaf vegetable with a pungent, yet slightly peppery flavor. It is typically sold in bouquets and can be refrigerated for up to five days if the stems are kept in water. Watercress is often used to garnish sandwiches, salads, soups, and stir-fries.

Figure 3-6. Leaf vegetables, also known as greens, are plant leaves that are often accompanied by edible leafstalks and shoots.

Edible Flowers

Edible flowers are the flowers of nonwoody plants that are prepared as vegetables. Edible flowers can be eaten raw or cooked. Examples of edible flowers include artichokes, broccoli, cauliflower, and squash blossoms. **See Figure 3-7.**

Production Tip

Edible flowers can be used to add color and flavor. For example, roses or lavender can be used to add color and sweetness to salads and desserts. Nasturtiums have a peppery flavor similar to watercress, and their buds can be substituted for capers.

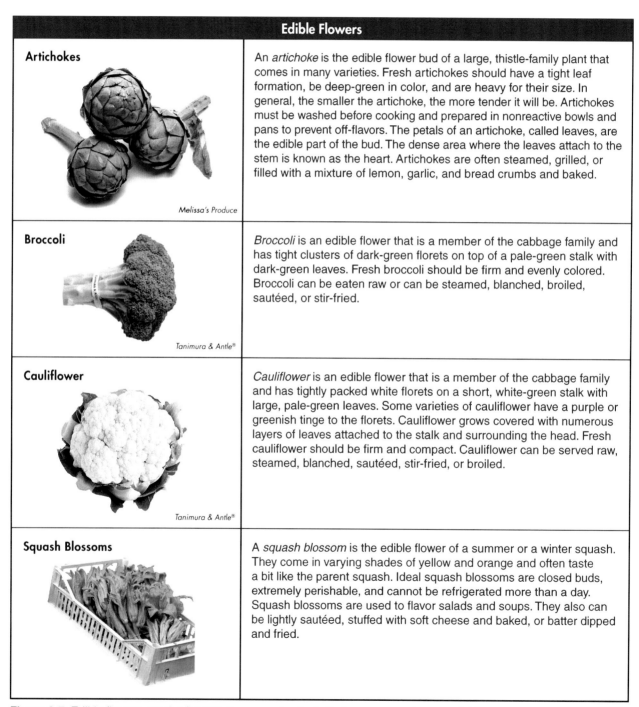

Edible Flowers	
Artichokes *Melissa's Produce*	An *artichoke* is the edible flower bud of a large, thistle-family plant that comes in many varieties. Fresh artichokes should have a tight leaf formation, be deep-green in color, and are heavy for their size. In general, the smaller the artichoke, the more tender it will be. Artichokes must be washed before cooking and prepared in nonreactive bowls and pans to prevent off-flavors. The petals of an artichoke, called leaves, are the edible part of the bud. The dense area where the leaves attach to the stem is known as the heart. Artichokes are often steamed, grilled, or filled with a mixture of lemon, garlic, and bread crumbs and baked.
Broccoli *Tanimura & Antle®*	*Broccoli* is an edible flower that is a member of the cabbage family and has tight clusters of dark-green florets on top of a pale-green stalk with dark-green leaves. Fresh broccoli should be firm and evenly colored. Broccoli can be eaten raw or can be steamed, blanched, broiled, sautéed, or stir-fried.
Cauliflower *Tanimura & Antle®*	*Cauliflower* is an edible flower that is a member of the cabbage family and has tightly packed white florets on a short, white-green stalk with large, pale-green leaves. Some varieties of cauliflower have a purple or greenish tinge to the florets. Cauliflower grows covered with numerous layers of leaves attached to the stalk and surrounding the head. Fresh cauliflower should be firm and compact. Cauliflower can be served raw, steamed, blanched, sautéed, stir-fried, or broiled.
Squash Blossoms	A *squash blossom* is the edible flower of a summer or a winter squash. They come in varying shades of yellow and orange and often taste a bit like the parent squash. Ideal squash blossoms are closed buds, extremely perishable, and cannot be refrigerated more than a day. Squash blossoms are used to flavor salads and soups. They also can be lightly sautéed, stuffed with soft cheese and baked, or batter dipped and fried.

Figure 3-7. Edible flowers are the flowers of nonwoody plants that are prepared as vegetables.

Edible Seeds

Edible seeds include some of the oldest recorded forms of food. Many edible seeds can be eaten raw, and all of them can be cooked. Examples of edible seeds include all varieties of legumes and sprouts. A *legume* is the edible seed of a nonwoody plant and grows in multiples within a pod. In some cases, the pods are eaten along with the seeds. Legumes are rich in fiber and protein and contain little or no fat. There are thousands of varieties of legumes, including beans, peas, pulses, and lentils.

Beans. A *bean* is the edible seed of various plants in the legume family. Beans are usually kidney-shaped or round and can be purchased fresh, canned, frozen, or dried. Popular varieties of beans include limas, cannellinis, anasazis, peruanos, calypsos, flageolets, pintos, kidney beans, great northern beans, and black beans. **See Figure 3-8.** Beans are often used to make hearty soups and are also served as sides. Some beans, such as pintos, can be puréed to make refried beans. Lima beans can be used to make succotash.

Most fresh beans can be eaten raw, steamed, sautéed, grilled, or fried. Some fresh bean varieties are called edible pods, meaning that both the exterior skin and the interior seeds are edible. For example, fresh green beans and fresh wax beans are actually immature beans with underdeveloped pods that are therefore edible. **See Figure 3-9.**

Beans with Edible Pods

Green Beans   **Wax Beans**

Melissa's Produce

Figure 3-9. Green beans and wax beans are beans with edible pods.

Beans

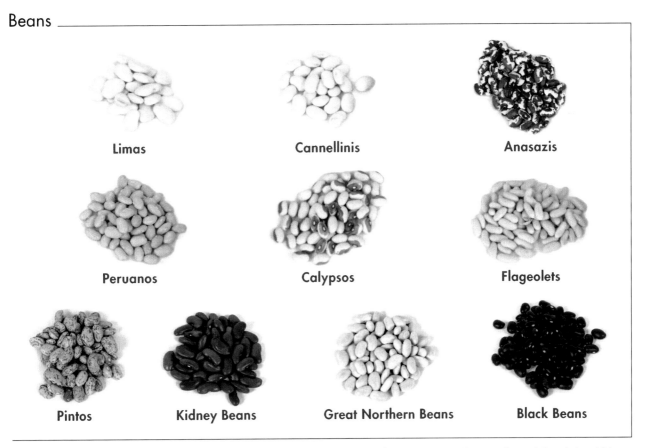

Figure 3-8. Popular varieties of beans include limas, cannellinis, anasazis, peruanos, calypsos, flageolets, pintos, kidney beans, great northern beans, and black beans.

Other varieties of fresh beans, such as edamame, do not have edible pods. *Edamame* are green soybeans housed within a fibrous, inedible pod. **See Figure 3-10.**

Edamame

Figure 3-10. Edamame are green soybeans housed within a fibrous, inedible pod.

Peas. A *pea* is the edible seed of various plants in the legume family. Most peas are round in shape. Some peas, called split peas, are harvested fully mature, left to dry, and then split. Some varieties of peas have edible pods. For example, snow peas have a flat pod that is entirely edible. Sugar snap peas have a more rounded pod, but are still tender enough to eat. **See Figure 3-11.** These edible pods can be eaten raw, steamed, sautéed, or fried.

Peas with Edible Pods

Snow Peas   Sugar Snap Peas

Melissa's Produce

Figure 3-11. Edible pods can be eaten raw, steamed, sautéed, or fried.

Pulses. A *pulse* is a dried seed of a legume. Dried beans and peas, such as cannellini beans and black-eyed peas, are shelled and then left to dry until they become rock hard. Pulses must be rehydrated by soaking them overnight or by using a quick-soaking process. Rehydration decreases the total cooking time and results in an even texture throughout. **See Figure 3-12.**

Pulses

Indian Harvest Specialtifoods, Inc./Rob Yuretich

Figure 3-12. Pulses are the dried seeds of legumes and must be rehydrated before being cooked.

Lentils. A *lentil* is a very small, dried pulse that has been split in half. There are many varieties of lentils, with colors ranging from white to green. **See Figure 3-13.** Unlike dried beans and peas, lentils do not have to be soaked. However, lentils must be thoroughly washed before cooking because they often contain small stones. Lentils are used to make soups, added to salads, combined and served with other vegetables, and served as sides. Lentils turn mushy when overcooked.

Lentils

Split White   Petite Crimson

Black Beluga   French Green

Figure 3-13. Varieties of lentils range in color from white to green.

Lentils

Nutrition Note

Lentils contain calcium and vitamins A and B. They are also a good source of iron and phosphorus.

Fruit-Vegetables

A *fruit-vegetable* is a botanical fruit that is sold, prepared, and served as a vegetable. Fruit-vegetables are typically more tart than sweet. Fruit-vegetables used in the professional kitchen include bell peppers, chiles, cucumbers, eggplants, okra, pumpkins, summer squashes, sweet corn, tomatillos, tomatoes, and winter squashes. **See Figure 3-14.**

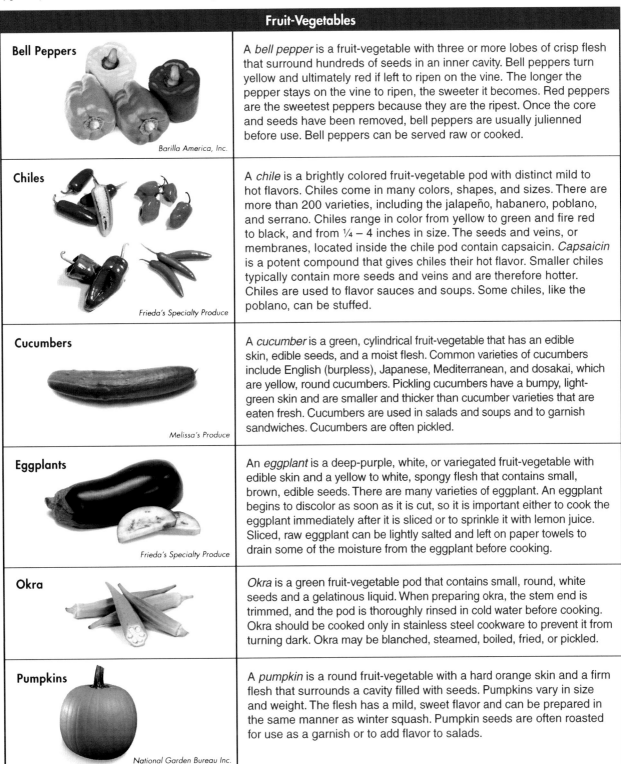

Fruit-Vegetables	
Bell Peppers *Barilla America, Inc.*	A *bell pepper* is a fruit-vegetable with three or more lobes of crisp flesh that surround hundreds of seeds in an inner cavity. Bell peppers turn yellow and ultimately red if left to ripen on the vine. The longer the pepper stays on the vine to ripen, the sweeter it becomes. Red peppers are the sweetest peppers because they are the ripest. Once the core and seeds have been removed, bell peppers are usually julienned before use. Bell peppers can be served raw or cooked.
Chiles *Frieda's Specialty Produce*	A *chile* is a brightly colored fruit-vegetable pod with distinct mild to hot flavors. Chiles come in many colors, shapes, and sizes. There are more than 200 varieties, including the jalapeño, habanero, poblano, and serrano. Chiles range in color from yellow to green and fire red to black, and from ¼ – 4 inches in size. The seeds and veins, or membranes, located inside the chile pod contain capsaicin. *Capsaicin* is a potent compound that gives chiles their hot flavor. Smaller chiles typically contain more seeds and veins and are therefore hotter. Chiles are used to flavor sauces and soups. Some chiles, like the poblano, can be stuffed.
Cucumbers *Melissa's Produce*	A *cucumber* is a green, cylindrical fruit-vegetable that has an edible skin, edible seeds, and a moist flesh. Common varieties of cucumbers include English (burpless), Japanese, Mediterranean, and dosakai, which are yellow, round cucumbers. Pickling cucumbers have a bumpy, light-green skin and are smaller and thicker than cucumber varieties that are eaten fresh. Cucumbers are used in salads and soups and to garnish sandwiches. Cucumbers are often pickled.
Eggplants *Frieda's Specialty Produce*	An *eggplant* is a deep-purple, white, or variegated fruit-vegetable with edible skin and a yellow to white, spongy flesh that contains small, brown, edible seeds. There are many varieties of eggplant. An eggplant begins to discolor as soon as it is cut, so it is important either to cook the eggplant immediately after it is sliced or to sprinkle it with lemon juice. Sliced, raw eggplant can be lightly salted and left on paper towels to drain some of the moisture from the eggplant before cooking.
Okra	*Okra* is a green fruit-vegetable pod that contains small, round, white seeds and a gelatinous liquid. When preparing okra, the stem end is trimmed, and the pod is thoroughly rinsed in cold water before cooking. Okra should be cooked only in stainless steel cookware to prevent it from turning dark. Okra may be blanched, steamed, boiled, fried, or pickled.
Pumpkins *National Garden Bureau Inc.*	A *pumpkin* is a round fruit-vegetable with a hard orange skin and a firm flesh that surrounds a cavity filled with seeds. Pumpkins vary in size and weight. The flesh has a mild, sweet flavor and can be prepared in the same manner as winter squash. Pumpkin seeds are often roasted for use as a garnish or to add flavor to salads.

Figure 3-14. (continued on next page)

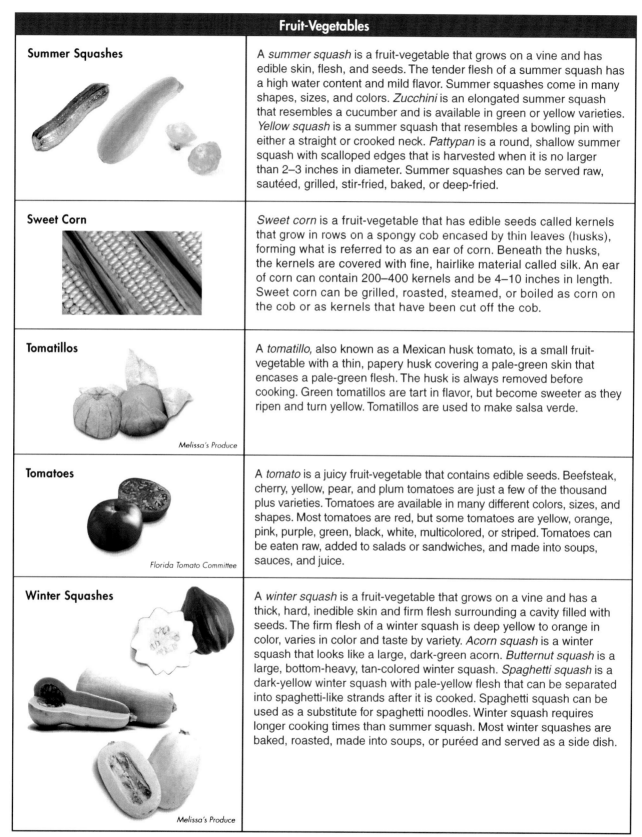

Fruit-Vegetables

Summer Squashes

A *summer squash* is a fruit-vegetable that grows on a vine and has edible skin, flesh, and seeds. The tender flesh of a summer squash has a high water content and mild flavor. Summer squashes come in many shapes, sizes, and colors. *Zucchini* is an elongated summer squash that resembles a cucumber and is available in green or yellow varieties. *Yellow squash* is a summer squash that resembles a bowling pin with either a straight or crooked neck. *Pattypan* is a round, shallow summer squash with scalloped edges that is harvested when it is no larger than 2–3 inches in diameter. Summer squashes can be served raw, sautéed, grilled, stir-fried, baked, or deep-fried.

Sweet Corn

Sweet corn is a fruit-vegetable that has edible seeds called kernels that grow in rows on a spongy cob encased by thin leaves (husks), forming what is referred to as an ear of corn. Beneath the husks, the kernels are covered with fine, hairlike material called silk. An ear of corn can contain 200–400 kernels and be 4–10 inches in length. Sweet corn can be grilled, roasted, steamed, or boiled as corn on the cob or as kernels that have been cut off the cob.

Tomatillos

A *tomatillo,* also known as a Mexican husk tomato, is a small fruit-vegetable with a thin, papery husk covering a pale-green skin that encases a pale-green flesh. The husk is always removed before cooking. Green tomatillos are tart in flavor, but become sweeter as they ripen and turn yellow. Tomatillos are used to make salsa verde.

Melissa's Produce

Tomatoes

A *tomato* is a juicy fruit-vegetable that contains edible seeds. Beefsteak, cherry, yellow, pear, and plum tomatoes are just a few of the thousand plus varieties. Tomatoes are available in many different colors, sizes, and shapes. Most tomatoes are red, but some tomatoes are yellow, orange, pink, purple, green, black, white, multicolored, or striped. Tomatoes can be eaten raw, added to salads or sandwiches, and made into soups, sauces, and juice.

Florida Tomato Committee

Winter Squashes

A *winter squash* is a fruit-vegetable that grows on a vine and has a thick, hard, inedible skin and firm flesh surrounding a cavity filled with seeds. The firm flesh of a winter squash is deep yellow to orange in color, varies in color and taste by variety. *Acorn squash* is a winter squash that looks like a large, dark-green acorn. *Butternut squash* is a large, bottom-heavy, tan-colored winter squash. *Spaghetti squash* is a dark-yellow winter squash with pale-yellow flesh that can be separated into spaghetti-like strands after it is cooked. Spaghetti squash can be used as a substitute for spaghetti noodles. Winter squash requires longer cooking times than summer squash. Most winter squashes are baked, roasted, made into soups, or puréed and served as a side dish.

Melissa's Produce

Figure 3-14. Fruit-vegetables are botanical fruits that are sold, prepared, and served as vegetables.

Sea Vegetables

Sea vegetables are edible saltwater plants that contain high amounts of dietary fiber, vitamins, and minerals. People living in coastal Asian countries, New Zealand, the Pacific Islands, coastal South American countries, Ireland, Scotland, Norway, and Iceland have been harvesting and eating sea vegetables for thousands of years.

Sea vegetables lend a salty flavor to food because of the minerals they absorb from the ocean. When adding sea vegetables to a dish, no salt should be added. Many sea vegetables also contain alginic acid, which is used as a stabilizer and a thickener when making processed foods such as ice creams, puddings, and pie fillings. Sea vegetables can be roasted along with other vegetables or crumbled and added to soups, sauces, salads, pastas, and rice dishes. Common varieties of sea vegetables used in the professional kitchen include dulse, kombu, nori, and wakame. **See Figure 3-15.**

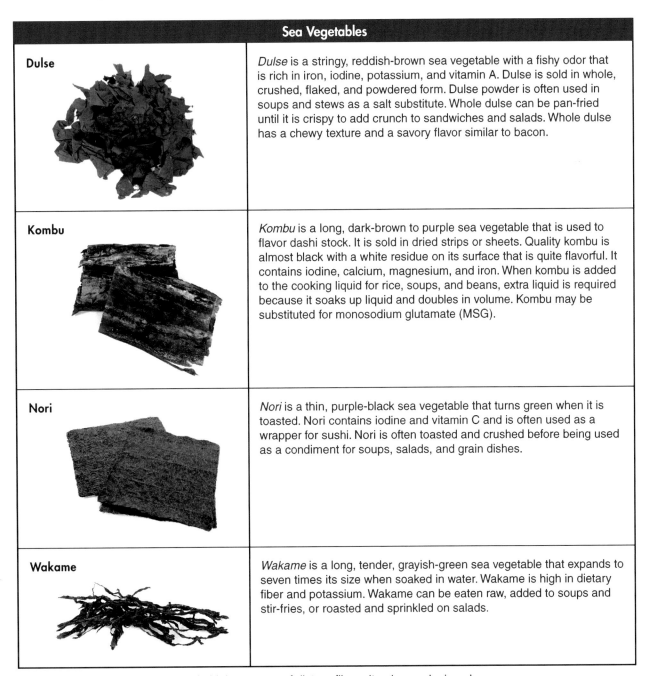

Sea Vegetables	
Dulse	*Dulse* is a stringy, reddish-brown sea vegetable with a fishy odor that is rich in iron, iodine, potassium, and vitamin A. Dulse is sold in whole, crushed, flaked, and powdered form. Dulse powder is often used in soups and stews as a salt substitute. Whole dulse can be pan-fried until it is crispy to add crunch to sandwiches and salads. Whole dulse has a chewy texture and a savory flavor similar to bacon.
Kombu	*Kombu* is a long, dark-brown to purple sea vegetable that is used to flavor dashi stock. It is sold in dried strips or sheets. Quality kombu is almost black with a white residue on its surface that is quite flavorful. It contains iodine, calcium, magnesium, and iron. When kombu is added to the cooking liquid for rice, soups, and beans, extra liquid is required because it soaks up liquid and doubles in volume. Kombu may be substituted for monosodium glutamate (MSG).
Nori	*Nori* is a thin, purple-black sea vegetable that turns green when it is toasted. Nori contains iodine and vitamin C and is often used as a wrapper for sushi. Nori is often toasted and crushed before being used as a condiment for soups, salads, and grain dishes.
Wakame	*Wakame* is a long, tender, grayish-green sea vegetable that expands to seven times its size when soaked in water. Wakame is high in dietary fiber and potassium. Wakame can be eaten raw, added to soups and stir-fries, or roasted and sprinkled on salads.

Figure 3-15. Sea vegetables contain high amounts of dietary fiber, vitamins, and minerals.

Mushrooms

Although mushrooms are not vegetables, they are prepared and used in the same manner as vegetables. A *mushroom* is the fleshy, spore-bearing body of an edible fungus that grows above the ground. Many mushrooms are commercially raised, but others are harvested from the wild. Both fresh and dried mushrooms are used in the professional kitchen.

Fresh mushrooms should be firm and not spotted or slimy. Fresh mushrooms must be stored in a cool, dry place. They also require air circulation to stay fresh, so storing them in paper bags works best. Fresh mushrooms should be cleaned with a damp towel and lightly rinsed, if necessary, before being used in a recipe. Dried mushrooms must be rehydrated before they are used in a recipe.

Courtesy of Chef Eric LeVine

Mushrooms are used to add flavor to many dishes and can be served raw, sautéed, stir-fried, or deep-fried. Common varieties of mushrooms used in the professional kitchen include button, chanterelle, enokitake, morel, oyster, porcini, portobello, shiitake, and wood ear mushrooms. **See Figure 3-16.**

Mushrooms	
Button Mushrooms *Mushroom Council*	A *button mushroom,* also known as a white mushroom, is a cultivated mushroom with a very smooth, rounded cap and completely closed gills atop a short stem. Button mushrooms are one of the most widely consumed mushrooms. Button mushrooms have a mild flavor that complements most foods.
Chanterelle Mushrooms *Mushroom Council*	A *chanterelle mushroom* is a trumpet-shaped mushroom that ranges in color from bright yellow to orange and has a nutty flavor and a chewy texture. Chanterelles tend to toughen when overcooked, so it is best to add them to a dish near the end of the cooking time.
Enokitake Mushrooms *Mushroom Council*	An *enokitake mushroom,* also known as an enoki or a snow puff mushroom, is a crisp, delicate mushroom that has spaghetti-like stems topped with white caps. Enokitake mushrooms have a crunchy texture and an almost fruity flavor. Fresh enokitake mushrooms should be wrapped in a paper towel and then a plastic bag before being refrigerated.
Morel Mushrooms *Melissa's Produce*	A *morel mushroom* is an uncultivated mushroom with a cone-shaped cap that ranges in height from 2–4 inches and in color from tan to very dark brown. Morels belong to the same fungus species as the truffle and are favored for their earthy, nutty flavor. Darker-colored morels tend to have a stronger flavor. Dried morels have a more intense and smoky flavor. Morel mushrooms are typically sautéed in butter.

Figure 3-16. (continued on next page)

Mushrooms

Mushrooms	
Oyster Mushrooms	An *oyster mushroom* is a broad, fanlike or oyster-shaped mushroom that varies in color from white to gray or tan to dark brown. Oyster mushrooms span 2–10 inches. Their flavor is mild and they have a slight odor that resembles anise. The firm white flesh varies in thickness. The gills of the mushroom are white to cream and descend toward the stalk, which is often not present at the point of sale. Oyster mushrooms may be sautéed or stir-fried.
Porcini Mushrooms	A *porcini mushroom,* also known as a cèpe, is an uncultivated, pale-brown mushroom with a smooth, meaty texture and a pungent flavor. Porcini mushrooms can weigh anywhere from 1 oz to 1 lb and have a cap that ranges from 1–10 inches in diameter. Dried porcini mushrooms must be softened in hot water for about 20 minutes before use. Porcinis can be substituted for cultivated mushrooms in most recipes.
Portobello Mushrooms *Mushroom Council*	A *portobello mushroom* is a very large and mature, brown cremini mushroom that has a flat cap measuring up to 6 inches in diameter. The gills of the portobello mushroom are fully exposed, leaving the mushroom without much moisture and creating a dense, meaty texture. Their woody stems are removed and used to flavor stocks and soups. The caps are typically used whole, but can be diced for use in a wide variety of dishes. Portobello mushrooms have a meaty, savory flavor and are often grilled for sandwiches or sliced and added to salads.
Shiitake Mushrooms	A *shiitake mushroom,* also known as a forest mushroom, is an amber, tan, brown, or dark-brown mushroom with an umbrella shape and curled edges. Shiitake mushroom caps range in size from 3–10 inches in diameter. Cooked shiitakes release a pinelike aroma and have a rich, earthy, savory flavor. The tough stems are usually removed and used to flavor stocks and soups. Shiitake mushrooms can be sautéed, broiled, or baked.
Wood Ear Mushroom	A *wood ear mushroom,* also known as a cloud ear or a tree ear mushroom, is a brownish-black, ear-shaped mushroom that has a slightly crunchy texture. When dried wood ear mushrooms are reconstituted they increase 5–6 times in size. The albino variety of wood ear mushrooms is white in color. Wood ear mushrooms have a delicate flavor and often absorb the taste of other ingredients in a dish.

Figure 3-16. An edible mushroom is the fleshy, spore-bearing body of an edible fungus that grows above the ground.

Purchasing Fresh Vegetables

Fresh vegetables are packed in cartons, lugs, flats, crates, or bushels and sold by weight or count. The weight or count of a packed container depends on the size and type of vegetable. Because most fresh vegetables have a short shelf life, it is important to know how to maximize their use. **See Figure 3-17.**

Fresh Vegetables

Vegetables

Barilla America, Inc.

Figure 3-17. Fresh vegetables are packed in cartons, lugs, flats, crates, or bushels and sold by weight or count.

The USDA voluntary grading system for fresh vegetables includes U.S. Extra Fancy, U.S. Fancy, U.S. Extra No. 1, U.S. No. 1, U.S. No. 2, and U.S. No. 3. The grade of vegetable to purchase depends on how the vegetable will be used. Recipes using fresh or slightly cooked vegetables require premium ingredients, while lesser grades are acceptable in soup recipes. Fresh vegetables are less expensive during their peak season.

Most vegetables should be stored in a produce cooler at a temperature of 41°F or below. Vegetables should always be stored away from fruits that emit ethylene gas, such as apples and bananas, as the gas can cause the vegetables to overripen and spoil. It is also important to store vegetables away from poultry, meat, seafood, and dairy products.

Vegetables such as potatoes, onions, garlic, and squash should be stored in a cool, dry location that is between 60°F and 70°F. Storing these vegetables in a refrigerator causes their starches to convert to sugars, which negatively alters their texture and flavor.

Canned Vegetables

Canned vegetables are a staple in the professional kitchen. Canned vegetables have already been cleaned, cut, peeled, cooked, and treated with heat to kill any harmful microorganisms. However, the canning process often softens vegetables and can sometimes cause nutrient loss. Canned vegetables are USDA graded as U.S. Grade A or U.S. Fancy, U.S. Grade B, and U.S. Grade C or U.S. Standard. Canned vegetables are packed by weight and sold in standard commercial sizes. **See Figure 3-18.**

Vegetable Can Sizes		
Can Size	**Weight**	**Cans Per Case**
No. 300	14–15 oz	36
No. 303	16–17 oz	36
No. 2	20 oz	24
No. 2½	28 oz	24
No. 5	46–51 oz	12
No. 10	6 lb 10 oz	6

Figure 3-18. Canned vegetables are a staple in the professional kitchen and come in standard sizes.

Canned vegetables can be stored for long periods if they are kept in a cool, dry place. Dented or bulging cans should be discarded as they may contain harmful bacteria. After opening canned vegetables, the unused portion should be placed in an airtight storage container, labeled, dated, and refrigerated.

Frozen Vegetables

Frozen vegetables offer the same convenience as canned vegetables, with an additional advantage. Frozen vegetables retain their color and nutrients better than canned vegetables. **See Figure 3-19.** The USDA grading system used for canned vegetables also applies to frozen vegetables. Frozen vegetables are usually packed in 1 lb or 2 lb bags. Some vegetables are individually quick-frozen to preserve their texture and appearance. Some vegetables are blanched before being frozen, which reduces overall cooking time. Other frozen vegetables are already fully cooked and need only to be heated for service.

Frozen Vegetables

Figure 3-19. Frozen vegetables often retain their color and nutrients better than canned vegetables.

Dried Vegetables

Dried vegetables have had most of their moisture removed by a food dehydrator or the freeze drying process. **See Figure 3-20.** *Freeze drying* is the process of removing the water content from a food and replacing it with a gas. Freeze dried vegetables retain more color, texture, and shape than dehydrated vegetables.

Dried Vegetables

Frieda's Specialty Produce

Figure 3-20. Dried vegetables, such as sun-dried tomatoes, have had most of their moisture removed by the dehydration process.

Dried onions are commonly used for convenience. For example, 1 lb of dried onions can be used instead of 8 lb of fresh onions. For maximum shelf life, dried vegetables should be stored in an airtight container in a cool, dry place.

CHECKPOINT 3-1

1. Describe common root vegetables used in the professional kitchen.

2. Describe common bulb vegetables used in the professional kitchen.

3. Describe common tubers used in the professional kitchen.

4. Describe common stem vegetables used in the professional kitchen.

5. Describe common leaf vegetables used in the professional kitchen.

6. Describe edible flowers used in the professional kitchen.

7. Describe common seed vegetables used in the professional kitchen.

8. Describe common fruit-vegetables used in the professional kitchen.

9. Describe common sea vegetables used in the professional kitchen.

10. Describe common mushrooms used in the professional kitchen.

11. Describe factors to consider when purchasing fresh vegetables.

12. Explain the role of canned vegetables in the professional kitchen.

13. Explain the role of frozen vegetables in the professional kitchen.

14. Explain the role of dried vegetables in the professional kitchen.

COOKING VEGETABLES

Common methods for cooking vegetables include steaming, blanching, grilling, broiling, baking, roasting, sautéing, and frying. Vegetables should be cooked until they are just tender enough to be easily digested. At this stage of cooking, most vegetables retain the majority of their nutritional value, flavor, and color. Overcooked vegetables often lose their bright colors, may become mushy in texture, and lose nutrients because vitamins and minerals are destroyed by excess heat.

The addition of acidic or alkaline ingredients when cooking vegetables causes chemical reactions that affect the color and texture of the vegetables. Acids, such as lemon juice or vinegar, are often added to a cooking liquid to contribute flavor to the dish. An alkali such as baking soda can be added to tough vegetables to speed up the softening process. Adding acids or alkalis also alters the natural pigment present in a vegetable. **See Figure 3-21.** The types of natural pigments in vegetables include chlorophyll, carotenoids, and flavonoids.

- *Chlorophyll* is an organic pigment found in green vegetables. When an acid is added to the cooking liquid, green vegetables turn a drab olive color, but retain their naturally firm texture. If an alkali is added to the cooking liquid, green vegetables become brighter in color but mushy in texture.
- A *carotenoid* is an organic pigment found in orange or yellow vegetables. Acids have little to no effect on carotenoids. Alkalis do not affect the color of carotenoids, but do cause orange and yellow vegetables to become mushy.
- A *flavonoid* is an organic pigment found in purple, dark-red, and white vegetables. Acids cause purple and dark-red vegetables to turn bright red. In contrast, alkaline ingredients cause purple and dark-red vegetables to turn blue and white vegetables to turn yellow. Alkalis also cause purple, dark-red, and white vegetables to have a mushy texture.

Steaming Vegetables

Steaming is one of the quickest ways to cook vegetables and minimizes nutrient loss. Steamed vegetables also retain much of their color and most of their firm texture. Large quantities of fresh or frozen vegetables can be steamed in a pressurized or convection steamer. **See Figure 3-22.** Steamed vegetables may also be sautéed.

Steaming Vegetables

Figure 3-22. Large quantities of fresh or frozen vegetables can be steamed in a pressurized or convection steamer.

Acid and Alkali Reactions				
Pigment	**Cooked Vegetables***		**Acid Added**	**Alkali Added**
Chlorophyll	Broccoli		Color loss	Mushy texture
Carotenoids	Carrots		Little or no effect	Mushy texture
Flavonoids	Beets		Brighter red	Turns blue; mushy texture

* No acidic or alkaline ingredients used

Figure 3-21. The addition of acidic or alkaline ingredients when cooking vegetables causes chemical reactions that affect the color and texture of the vegetables.

Irinox USA

Blanching Vegetables

Blanching is a moist-heat cooking technique in which food is briefly parcooked and then shocked by placing it in ice-cold water to stop the cooking process. Fresh vegetables are often blanched to highlight their bright color before they are finished with another cooking technique. For example, vegetables such as asparagus are often blanched and then finished in a broiler or on a grill at time of service. Other vegetables, such as tomatoes, may be blanched to make it easier to remove their skin.

Nutrition Note

Blanching not only brightens the colors of some fruits and vegetables, but also allows foods to retain their vitamins and minerals.

Grilling and Broiling Vegetables

Grilling and broiling are both fast and easy ways to prepare vegetables. Fresh vegetables are seasoned as desired, drizzled with a little oil, and then placed under the broiler or directly on the grill. The size of the vegetables determines how long they need to cook. Grilling and broiling caramelize the sugars in vegetables, which results in a sweeter flavor. **See Figure 3-23.**

Grilled Vegetables

Tanimura & Antle®

Figure 3-23. Grilling vegetables caramelizes their sugars, giving them a sweeter flavor.

Procedure for Blanching

1. Bring a pot of water to a boil and prepare a separate ice bath. *Note:* Adding salt to the boiling water when blanching green vegetables intensifies their color.

2. Place the cleaned and prepared items in the rapidly boiling water until the desired result is achieved.

3. Remove the items and immediately submerge them in the ice bath to stop the cooking process.

Baking and Roasting Vegetables

Baking and roasting are excellent ways to prepare vegetables. Dishes such as a broccoli and cheese casserole are typically baked. Many root vegetables such as onions, carrots, turnips, parsnips, and various types of potatoes are often roasted together or placed alongside a large cut of meat. When baking and roasting vegetables, it is important to cut vegetables into uniform sizes to ensure doneness. Some vegetables, such as peppers, can be fire-roasted. Fire-roasting a vegetable over an open flame allows the item to be cooked whole.

Sautéing Vegetables

Vegetables prepared for sautéing or stir-frying are usually diced small or thinly sliced. Fresh vegetables are sautéed and stir-fried very quickly in a hot pan with a small amount of oil. **See Figure 3-24.** The finished vegetables are firm.

Sautéed and Stir-Fried Vegetables

National Honey Board

Sautéed

U.S. Apple Association

Stir-Fried

Figure 3-24. Vegetables are sautéed and stir-fried very quickly in a hot pan with a small amount of oil.

Frying Vegetables

Deep-fried vegetables such as onions, mushrooms, cauliflower, zucchini, and eggplant are popular appetizers. **See Figure 3-25.** These vegetables are batter-coated and fried until crisp. French fries are also a popular fried vegetable.

Fried Vegetables

Photo Courtesy of McCain Foods USA

Figure 3-25. Deep-fried vegetables such as onions rings are popular appetizers.

Nutrition Note

When frying vegetables, it is difficult to calculate the nutritional value of the frying oil. The nutritional value varies depending on cook time and temperature, ingredient density, and type of oil used. To compensate, a 10% retention rate is typically used to calculate the nutritional value of the oil.

Plating Vegetables

Vegetables add color and texture to a plate. The conventional clock system of placing vegetables at 2 o'clock, proteins at 6 o'clock, and starches at 11 o'clock is a good method to follow when plating. Vegetables should cover about half of the plate, while the protein should cover a fourth and the starch the other fourth. Green vegetables should be complemented with vegetables of other colors, such as carrots, bell peppers, or beets. Also, it is important to avoid clustering the vegetables. This is so that the contrasting colors can be seen. Different shapes and cuts of vegetables also add visual interest. Contrasting hard with soft vegetables and smooth with coarse vegetables will enhance the presentation. **See Figure 3-26.**

Plating Vegetables

Daniel NYC

Figure 3-26. Contrasting hard with soft vegetables and smooth with coarse vegetables will enhance the plate presentation.

CHECKPOINT 3-2

1. Explain how acidic and alkaline ingredients affect cooked vegetables.

2. Identify eight methods that can be used to cook vegetables.

3. Steam or blanch a vegetable and evaluate the quality of the prepared dish.

4. Grill or broil a vegetable and evaluate the quality of the prepared dish.

5. Bake or roast a vegetable and evaluate the quality of the prepared dish.

6. Explain how to fire-roast peppers.

7. Fire-roast a vegetable and evaluate the quality of the prepared dish.

8. Sauté a vegetable and evaluate the quality of the prepared dish.

9. Stir-fry a vegetable and evaluate the quality of the prepared dish.

10. Deep-fry a vegetable and evaluate the quality of the prepared dish.

FRUIT CLASSIFICATIONS

A *fruit* is the edible, ripened ovary of a flowering plant that usually contains one or more seeds. The vast assortment of available fruits is primarily due to the large number of fruit varieties and hybrid fruits created by fruit growers over the last 2000 years. **See Figure 3-27.** A *variety fruit* is a fruit that is the result of breeding two or more fruits of the same species that have different characteristics. For example, a Jonagold apple is a variety of apple created by breeding a Jonathan apple with a Golden Delicious apple. A *hybrid fruit* is a fruit that is the result of crossbreeding two or more fruits of different species to obtain a completely new fruit. For example, the loganberry was created by crossbreeding a raspberry and a blackberry.

Fruit Varieties

Figure 3-27. The vast assortment of available fruits is primarily due to the large number of fruit varieties and hybrid fruits created by fruit growers over the last 2000 years.

Fruits are nutritious because they are high in water, dietary fiber, vitamins, fructose, and antioxidants. Fruit adds color, texture, and flavor to a meal. Fruit is classified into several major categories, including berries, grapes, pomes, drupes, melons, citrus fruits, tropical fruits, exotic fruits, and fruit-vegetables.

Berries

A *berry* is a type of fruit that is small and has many tiny, edible seeds. Quality berries are sweet and evenly colored. Some berries are actually aggregate fruits. An *aggregate fruit* is a cluster of very tiny fruits. Blackberries, boysenberries, loganberries, and raspberries are aggregate fruits. Berries are harvested ripe because they do not continue to ripen after harvest. Berries commonly used in the professional kitchen include blackberries, blueberries, boysenberries, cranberries, currants, gooseberries, loganberries, raspberries, and strawberries. **See Figure 3-28.**

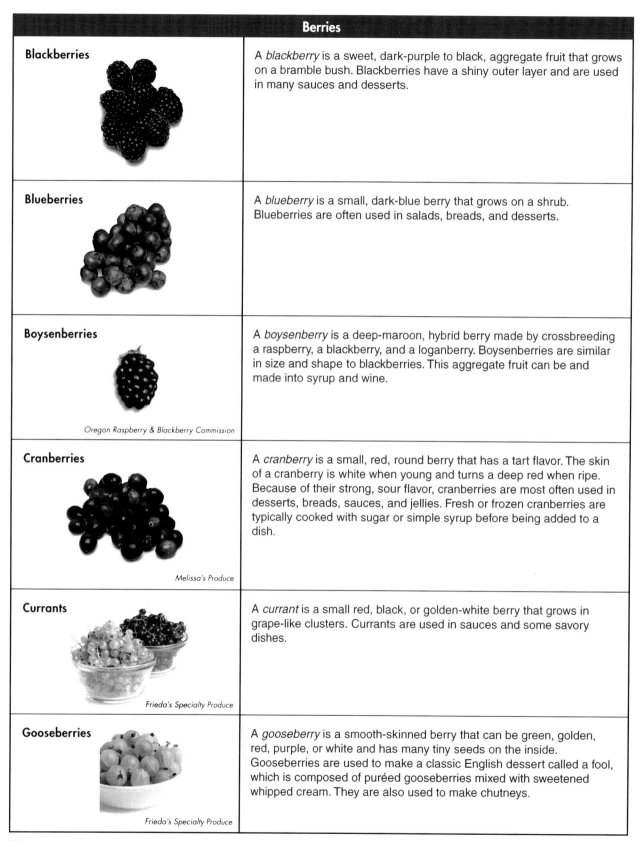

Berries	
Blackberries	A *blackberry* is a sweet, dark-purple to black, aggregate fruit that grows on a bramble bush. Blackberries have a shiny outer layer and are used in many sauces and desserts.
Blueberries	A *blueberry* is a small, dark-blue berry that grows on a shrub. Blueberries are often used in salads, breads, and desserts.
Boysenberries *Oregon Raspberry & Blackberry Commission*	A *boysenberry* is a deep-maroon, hybrid berry made by crossbreeding a raspberry, a blackberry, and a loganberry. Boysenberries are similar in size and shape to blackberries. This aggregate fruit can be and made into syrup and wine.
Cranberries *Melissa's Produce*	A *cranberry* is a small, red, round berry that has a tart flavor. The skin of a cranberry is white when young and turns a deep red when ripe. Because of their strong, sour flavor, cranberries are most often used in desserts, breads, sauces, and jellies. Fresh or frozen cranberries are typically cooked with sugar or simple syrup before being added to a dish.
Currants *Frieda's Specialty Produce*	A *currant* is a small red, black, or golden-white berry that grows in grape-like clusters. Currants are used in sauces and some savory dishes.
Gooseberries *Frieda's Specialty Produce*	A *gooseberry* is a smooth-skinned berry that can be green, golden, red, purple, or white and has many tiny seeds on the inside. Gooseberries are used to make a classic English dessert called a fool, which is composed of puréed gooseberries mixed with sweetened whipped cream. They are also used to make chutneys.

Figure 3-28. (continued on next page)

Berries	
Loganberries *Oregon Raspberry & Blackberry Commission*	A *loganberry* is a red-purple hybrid berry made by crossbreeding a raspberry and a blackberry. Loganberries are similar in size and shape to blackberries but are redder in color. Loganberries can be made into a syrup.
Raspberries	A *raspberry* is a slightly tart, red aggregate fruit that grows in clusters. The velvety soft texture of raspberries makes them one of the most fragile fruits. Raspberries are used in a variety of desserts and in sweet sauces to complement desserts, as well as in glazes for savory dishes such as roasted meats or poultry.
Strawberries *California Strawberry Commission*	A *strawberry* is a bright-red, heart-shaped berry covered with tiny black seeds. A ripe strawberry should be evenly red and free of brown or soft spots. Strawberries are widely used in desserts, sauces, and salads.

Figure 3-28. A berry is a small fruit with many tiny, edible seeds.

Grapes

A *grape* is an oval fruit that has a smooth skin and grows on woody vines in large clusters. Grapes are the most widely grown fruit because of their use in winemaking. Table grapes are the varieties suitable for eating, as opposed to varieties grown specifically for winemaking. The two main classifications of table grapes are white grapes, which are usually green in color, and black grapes, which are usually red to dark blue in color. Most of the flavor of a grape comes from its skin. It is important to choose grapes that are firm with no discoloration. Table grapes frequently used in the professional kitchen include Thompson, red flame, and Concord grapes. **See Figure 3-29.** Dried grapes are called raisins.

Thompson Grapes. A *Thompson grape* is a seedless grape that is pale to light green in color. Thompson grapes may be used in salads or served with a cheese platter. Dried seedless grapes are called sultanas.

Red Flame Grapes. A *red flame grape* is a seedless grape that ranges from a light purple-red color to a dark-purple color. Red flame grapes are typically sweeter and crisper in texture than Thompson grapes. Red flame grapes are often used as an accompaniment to salads, cheese trays, or platters.

Grapes

Figure 3-29. Common varieties of table grapes include Concord, red flame, and Thompson grapes.

Concord Grapes. A *Concord grape* is a seeded grape with a deep black color. It is sometimes called a slipskin grape, as its skin is very easily separated from the fruit. Concord grapes are available in the market as a table grape but, because they have seeds, they are used less often than the Thompson or red flame varieties.

Pomes

A *pome* is a fleshy fruit that contains a core of seeds and has an edible skin. Pomes have thin skin, grow on trees or bushes, and are an excellent source of antioxidants and dietary fiber. Quality pomes are free of blemishes and bruises and have no soft spots. Pomes often used in the professional kitchen include apples, pears, and quinces. **See Figure 3-30.**

Pomes

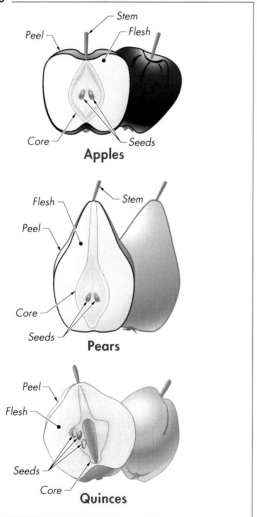

Figure 3-30. A pome is a fleshy fruit that contains a core of seeds and an edible peel. Apples, pears, and quinces are pomes.

Apples. An *apple* is a hard, round pome that can range in flavor from sweet to tart and in color from pale yellow to dark red. There are more than 7500 cultivated varieties of apples. Some varieties are not suitable for cooking purposes, and others become bitter when baked. In general, tart apples, or apples that have a high acid content, are desirable for cooking because of their intense flavor. **See Figure 3-31.**

Common Apple Varieties		
Name	**Description**	**Availability**
Cortland	Sweet, hint of tartness, juicy tender, white flesh	Sept. to Apr.
Crispin	Sweet, juicy, crisp	Oct. to Sept.
Empire	Sweet, tart, juicy, creamy-white flesh	Sept. to July
Fuji	Spicy, sweet, juicy, firm cream-colored flesh, tender skin	Oct. to June
Gala	Yellow to red, sweet, juicy, crisp yellow flesh	Sept. to June
Golden Delicious	Sweet, crisp, light yellow flesh	Sept. to June
Granny Smith	Tart, crisp, juicy	Sept. to June
Idared	Sweet, tart, juicy, firm, pale yellow-green or rosy pink flesh	Oct. to Aug.
Jonagold	Tangy, sweet	Oct. to May
McIntosh	Sweet, tangy, juicy, tender, white flesh	Sept. to June
Rome	Mildly tart, firm, greenish-white flesh	Oct. to Sept.

U.S. Apple Association

Figure 3-31. Common varieties of apples are used to prepare a vast array of dishes, including salads, sauces, purées, desserts, and savory dishes.

Pears. A *pear* is a bell-shaped pome with a thin peel and sweet flesh. Pears should be harvested before they are ripe, but if picked too early they will not develop their full flavor. If left to fully ripen on the tree, pears develop concentrations of cellulose, resulting in a grainy texture. There are thousands of different kinds of pears. Common varieties of pears include Anjou, Asian, Bartlett, Bosc, Comice, and Seckel pears. **See Figure 3-32.**

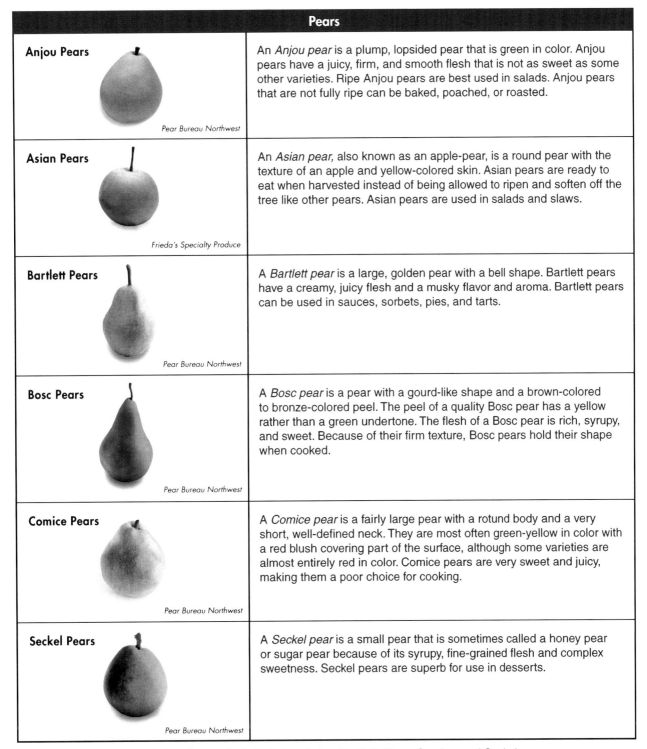

Pears	
Anjou Pears *Pear Bureau Northwest*	An *Anjou pear* is a plump, lopsided pear that is green in color. Anjou pears have a juicy, firm, and smooth flesh that is not as sweet as some other varieties. Ripe Anjou pears are best used in salads. Anjou pears that are not fully ripe can be baked, poached, or roasted.
Asian Pears *Frieda's Specialty Produce*	An *Asian pear,* also known as an apple-pear, is a round pear with the texture of an apple and yellow-colored skin. Asian pears are ready to eat when harvested instead of being allowed to ripen and soften off the tree like other pears. Asian pears are used in salads and slaws.
Bartlett Pears *Pear Bureau Northwest*	A *Bartlett pear* is a large, golden pear with a bell shape. Bartlett pears have a creamy, juicy flesh and a musky flavor and aroma. Bartlett pears can be used in sauces, sorbets, pies, and tarts.
Bosc Pears *Pear Bureau Northwest*	A *Bosc pear* is a pear with a gourd-like shape and a brown-colored to bronze-colored peel. The peel of a quality Bosc pear has a yellow rather than a green undertone. The flesh of a Bosc pear is rich, syrupy, and sweet. Because of their firm texture, Bosc pears hold their shape when cooked.
Comice Pears *Pear Bureau Northwest*	A *Comice pear* is a fairly large pear with a rotund body and a very short, well-defined neck. They are most often green-yellow in color with a red blush covering part of the surface, although some varieties are almost entirely red in color. Comice pears are very sweet and juicy, making them a poor choice for cooking.
Seckel Pears *Pear Bureau Northwest*	A *Seckel pear* is a small pear that is sometimes called a honey pear or sugar pear because of its syrupy, fine-grained flesh and complex sweetness. Seckel pears are superb for use in desserts.

Figure 3-32. Common varieties of pears include Anjou, Asian, Bartlett, Bosc, Comice, and Seckel pears.

Quinces. A *quince* is a hard yellow pome that grows in warm climates. **See Figure 3-33.** Quinces are not eaten raw because they have a bitter taste. This characteristic disappears during cooking. Quinces are typically cooked in sugar syrup, which turns the fruit slightly darker with a pinkish tint.

Drupes

A *drupe,* also known as a stone fruit, is a type of fruit that contains one hard seed or pit. Drupes grow on shrubs and trees and are usually harvested before they are ripe. High-quality drupes are free of blemishes or bruises. Apricots, avocados, cherries, dates, nectarines, olives, peaches, and plums are all drupes that are used in the professional kitchen. **See Figure 3-34.**

Quinces

Frieda's Specialty Produce

Figure 3-33. Quinces are hard yellow pomes that grow in warm climates.

Drupes	
Apricots	An *apricot* is a drupe that has pale orange-yellow skin with a fine, downy texture and a sweet and aromatic flesh. Apricots are quite delicate and may be harvested before they are ripe to avoid damage in shipping. Apricots are a popular choice for fruit tarts, dessert sauces, pastry fillings, and savory sauces for meats or poultry.
Avocados	An *avocado*, also known as an alligator pear, is a pear-shaped drupe with a rough green skin and a large pit surrounded by yellow-green flesh. The inedible skin and pit must be removed before the avocado can be eaten. The color of the skin varies depending on the variety. The skin of a Hass avocado turns black when ripe. Other varieties remain green when ripe. Avocados are full of vitamins and minerals and contain both omega 3 and 6 fatty acids. The majority of the fat in avocados is monounsaturated. Avocados also are a good source of protein and do not contain cholesterol. Avocados are often made into guacamole, used in salads, as a sandwich spread, and as an accompaniment to savory dishes.
Cherries	A *cherry* is a small, smooth-skinned drupe that grows in a cluster on a cherry tree. Cherries have a long, thin stem that holds them on the tree. The skin of cherries typically ranges in color from a bright red to a deep red that is nearly black. There are also golden-skinned varieties. The flesh of a cherry is pulpy and juicy and ranges in color from a dark yellow-orange to a deep reddish black. Cherries are classified as sweet or sour. Sweet cherries include Bing cherries, Gean cherries, and Rainier cherries. Sour cherries include Montmorency cherries and Morello cherries. Cherries are often used to make sauces, chutneys, pies, and a variety of desserts.
National Cherry Growers and Industries Foundation	

Figure 3-34. (continued on next page)

Drupes	
Dates *Frieda's Specialty Produce*	A *date* is a plump, juicy, and meaty drupe that grows on a date palm tree. Dates are high in dietary fiber, carbohydrates, and potassium. They are also low in fat, cholesterol, and sodium. Dates may be dried or stuffed with savory or sweet fillings after the pit has been removed. Dates can also be candied.
Nectarines	A *nectarine* is a sweet, slightly tart, orange to yellow drupe with a firm, yellow flesh and a large oval pit. Nectarines share many characteristics with peaches. Like peaches, nectarines need to ripen at room temperature to prevent them from becoming mealy. Virtually any recipe that calls for peaches can also be made with nectarines.
Olives	An *olive* is a small, green or black drupe that is grown for both the fruit and its oil. Unripe olives are green. Olives that are tree-ripened naturally turn dark brown to black. Fresh olives are bitter and their flavor reflects how ripe they were at harvest and how they were processed. Most olives are brined or salt cured and then packed in olive oil or vinegar. Black olives, also known as Mission olives, get their color and flavor from lye curing and oxygenation. Greek Kalamata olives and French Niçoise olives are also popular varieties. Pitted and unpitted olives are used in both raw and cooked dishes.
Peaches	A *peach* is a sweet, orange to yellow drupe with downy skin. The flesh of a peach is juicy, yet firm enough to hold its shape. The skin is edible, but the large oval pit inside the peach is not. Peaches should be ripened at room temperature and then refrigerated to prevent them from becoming mealy in texture. Peaches are used in salsas, chutneys, pies, and pastries.
Plums	A *plum* is an oval-shaped drupe that grows on trees in warm climates and comes in a variety of colors such as blue-purple, red, yellow, or green. There are more than 2000 varieties of plums. Some types of plums are sweet, juicy, and fragrant, while others can be sour, crisp, or mealy. The flesh can be red-orange, yellow, or green-yellow. Plums can be baked in cobblers and tarts. Dried plums are known as prunes.

Figure 3-34. A drupe, also known as a stone fruit, contains one hard seed or pit.

Melons

A *melon* is a type of fruit that has a hard outer rind (skin) and a soft inner flesh that contains many seeds. The hard outer rind can be netted, ribbed, or smooth in texture. Most melons are picked just before they are ripe. Ripe melons are firm and have a good aroma. Melons are a good source of vitamin C and potassium and are relatively low in calories.

With the exception of watermelon, the seeds are typically removed from fresh melons before service. The rind may or may not be removed depending on the desired presentation. Melons can be served as sides, used as garnishes, or made into cold soups, sorbets, ice creams, or parfaits. Canary melons, cantaloupes, casaba melons, Crenshaw melons, honeydew melons, muskmelons, and watermelons are all used in the professional kitchen. **See Figure 3-35.**

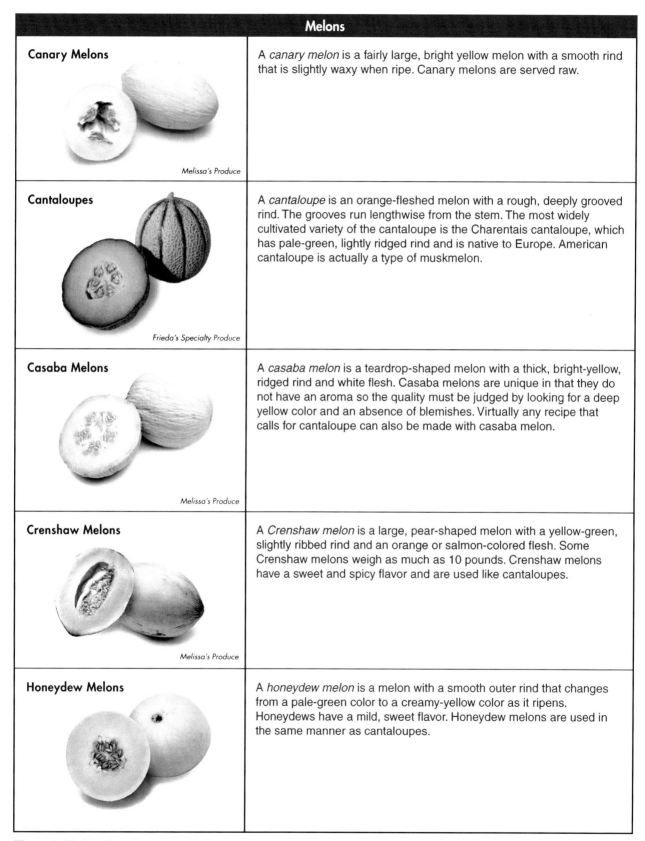

Melons

Canary Melons

Melissa's Produce

A *canary melon* is a fairly large, bright yellow melon with a smooth rind that is slightly waxy when ripe. Canary melons are served raw.

Cantaloupes

Frieda's Specialty Produce

A *cantaloupe* is an orange-fleshed melon with a rough, deeply grooved rind. The grooves run lengthwise from the stem. The most widely cultivated variety of the cantaloupe is the Charentais cantaloupe, which has pale-green, lightly ridged rind and is native to Europe. American cantaloupe is actually a type of muskmelon.

Casaba Melons

Melissa's Produce

A *casaba melon* is a teardrop-shaped melon with a thick, bright-yellow, ridged rind and white flesh. Casaba melons are unique in that they do not have an aroma so the quality must be judged by looking for a deep yellow color and an absence of blemishes. Virtually any recipe that calls for cantaloupe can also be made with casaba melon.

Crenshaw Melons

Melissa's Produce

A *Crenshaw melon* is a large, pear-shaped melon with a yellow-green, slightly ribbed rind and an orange or salmon-colored flesh. Some Crenshaw melons weigh as much as 10 pounds. Crenshaw melons have a sweet and spicy flavor and are used like cantaloupes.

Honeydew Melons

A *honeydew melon* is a melon with a smooth outer rind that changes from a pale-green color to a creamy-yellow color as it ripens. Honeydews have a mild, sweet flavor. Honeydew melons are used in the same manner as cantaloupes.

Figure 3-35. (continued on next page)

Melons	
Muskmelons 	A *muskmelon* is a round, orange-fleshed melon with a beige or brown, netted rind. The inside flesh can range in color from salmon to an orange-yellow color. There are many hybrids derived from muskmelon, such as the American cantaloupe. A *Santa Claus melon* is a large, mottled yellow and green variety of muskmelon that has a slightly waxy skin and soft stem end when ripe. A *Persian melon,* also known as a patelquat, is a green muskmelon with a finely textured net on the rind. The green coloring lightens and the netting turns brown as the melon ripens. Muskmelons are typically sliced and served raw.
Watermelons *National Watermelon Promotion Board*	A *watermelon* is a sweet, extremely juicy melon that is round or oblong in shape, with pink, red, or golden flesh and green skin. The watermelon is named for its high water content—watermelons are over 90% water. The weight of a watermelon varies by variety, but can weigh up to 30 pounds. The thick rind ranges from light green to dark green in color and is often striped or solid. Some watermelons are seedless while others contain a lot of seeds. Watermelon seeds thicken and turn black when mature. Immature watermelon seeds are thin and white. Watermelon is most often enjoyed fresh or puréed into sauces and cold soups.

Figure 3-35. Melons have a hard outer rind and a soft inner flesh that contains many seeds.

Citrus Fruits

A *citrus fruit* is a type of fruit with a brightly colored, thick rind and pulpy, segmented flesh that grows on trees in warm climates. The peel and the pith are both removed during preparation because of their bitter taste. The *peel* is the thick outer rind of a citrus fruit. The *pith* is the white layer just beneath the peel of a citrus fruit. Citrus fruits are harvested fully ripe, as they do not continue to ripen after they are picked. Fresh citrus should be kept refrigerated to extend storage life. Citrus fruits are an excellent source of vitamin C and are quite acidic. Common citrus fruits used in the professional kitchen include grapefruits, lemons, limes, mandarins, oranges, tangerines, and ugli fruit. **See Figure 3-36.**

Citrus Fruits	
Grapefruits	A *grapefruit* is a round citrus fruit with a thick, yellow outer rind and tart flesh. Common varieties of grapefruit include white grapefruit, pink grapefruit, and ruby red grapefruit. White grapefruit has pale-yellow flesh and pink grapefruit has pink-colored flesh. Ruby red grapefruit has red-colored flesh. Fresh grapefruit and grapefruit juice are commonly offered on breakfast menus.
Lemons	A *lemon* is a tart yellow citrus fruit with high acidity levels. The juice and rind of lemons are used in many dishes. Lemon juice is used in desserts, as well as many types of sauces that flavor poultry, fish, and shellfish. Lemon juice can also be used as a salad dressing or as a flavoring in marinades and beverages. A *Meyer lemon* is cross between a lemon and an orange. Meyer lemons are round and have a smooth, dark-yellow peel. The flesh has a yellow-orange color and the juice is sweeter and less acidic than a regular lemon.

Figure 3-36. (continued on next page)

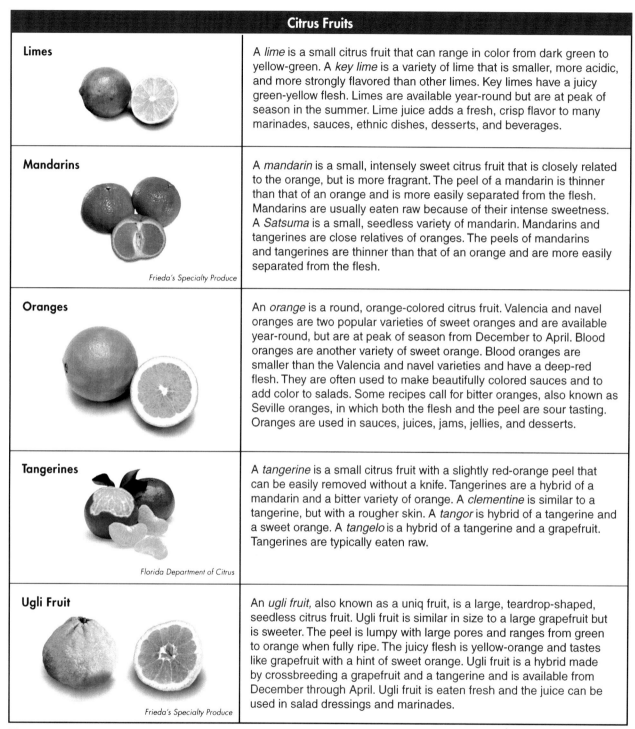

Citrus Fruits	
Limes	A *lime* is a small citrus fruit that can range in color from dark green to yellow-green. A *key lime* is a variety of lime that is smaller, more acidic, and more strongly flavored than other limes. Key limes have a juicy green-yellow flesh. Limes are available year-round but are at peak of season in the summer. Lime juice adds a fresh, crisp flavor to many marinades, sauces, ethnic dishes, desserts, and beverages.
Mandarins *Frieda's Specialty Produce*	A *mandarin* is a small, intensely sweet citrus fruit that is closely related to the orange, but is more fragrant. The peel of a mandarin is thinner than that of an orange and is more easily separated from the flesh. Mandarins are usually eaten raw because of their intense sweetness. A *Satsuma* is a small, seedless variety of mandarin. Mandarins and tangerines are close relatives of oranges. The peels of mandarins and tangerines are thinner than that of an orange and are more easily separated from the flesh.
Oranges	An *orange* is a round, orange-colored citrus fruit. Valencia and navel oranges are two popular varieties of sweet oranges and are available year-round, but are at peak of season from December to April. Blood oranges are another variety of sweet orange. Blood oranges are smaller than the Valencia and navel varieties and have a deep-red flesh. They are often used to make beautifully colored sauces and to add color to salads. Some recipes call for bitter oranges, also known as Seville oranges, in which both the flesh and the peel are sour tasting. Oranges are used in sauces, juices, jams, jellies, and desserts.
Tangerines *Florida Department of Citrus*	A *tangerine* is a small citrus fruit with a slightly red-orange peel that can be easily removed without a knife. Tangerines are a hybrid of a mandarin and a bitter variety of orange. A *clementine* is similar to a tangerine, but with a rougher skin. A *tangor* is hybrid of a tangerine and a sweet orange. A *tangelo* is a hybrid of a tangerine and a grapefruit. Tangerines are typically eaten raw.
Ugli Fruit *Frieda's Specialty Produce*	An *ugli fruit,* also known as a uniq fruit, is a large, teardrop-shaped, seedless citrus fruit. Ugli fruit is similar in size to a large grapefruit but is sweeter. The peel is lumpy with large pores and ranges from green to orange when fully ripe. The juicy flesh is yellow-orange and tastes like grapefruit with a hint of sweet orange. Ugli fruit is a hybrid made by crossbreeding a grapefruit and a tangerine and is available from December through April. Ugli fruit is eaten fresh and the juice can be used in salad dressings and marinades.

Figure 3-36. A citrus fruit has a brightly colored, thick rind and pulpy, segmented flesh.

Many recipes call for the zest of a citrus fruit, such as a lemon or a lime. *Zest* is the colored, outermost layer of the peel of a citrus fruit that contains a high concentration of oil. Zest is used to add flavor to a dish. Fine shreds of zest are made by gently rubbing the fruit against a zester or a rasp grater.

See Figure 3-37. Thin strips of zest can be obtained by running a five-hole zester along the surface of the rind. Large strips of zest can be obtained by using a vegetable peeler to peel off the colored surface of the rind and then be julienned. Zest may also be candied for use as a confection or a decoration.

Zesting

Browne-Halco (NJ)

Figure 3-37. Zesting is the process of removing the thin, colored peel of a citrus fruit.

Tropical Fruits

A *tropical fruit* is a type of fruit that comes from a hot, humid location but is readily available. Tropical fruits range in flavor from sweet to tangy and in texture from soft to crisp. Bananas, coconuts, figs, guavas, kiwifruit, mangoes, papayas, persimmons, pineapples, plantains, pomegranates, and prickly pears are tropical fruits that are commonly used in the professional kitchen. **See Figure 3-38.**

Frieda's Specialty Produce

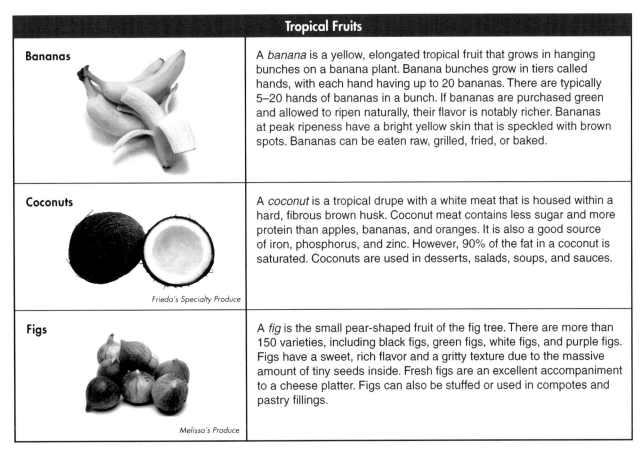

Tropical Fruits	
Bananas	A *banana* is a yellow, elongated tropical fruit that grows in hanging bunches on a banana plant. Banana bunches grow in tiers called hands, with each hand having up to 20 bananas. There are typically 5–20 hands of bananas in a bunch. If bananas are purchased green and allowed to ripen naturally, their flavor is notably richer. Bananas at peak ripeness have a bright yellow skin that is speckled with brown spots. Bananas can be eaten raw, grilled, fried, or baked.
Coconuts *Frieda's Specialty Produce*	A *coconut* is a tropical drupe with a white meat that is housed within a hard, fibrous brown husk. Coconut meat contains less sugar and more protein than apples, bananas, and oranges. It is also a good source of iron, phosphorus, and zinc. However, 90% of the fat in a coconut is saturated. Coconuts are used in desserts, salads, soups, and sauces.
Figs *Melissa's Produce*	A *fig* is the small pear-shaped fruit of the fig tree. There are more than 150 varieties, including black figs, green figs, white figs, and purple figs. Figs have a sweet, rich flavor and a gritty texture due to the massive amount of tiny seeds inside. Fresh figs are an excellent accompaniment to a cheese platter. Figs can also be stuffed or used in compotes and pastry fillings.

Figure 3-38. (continued on next page)

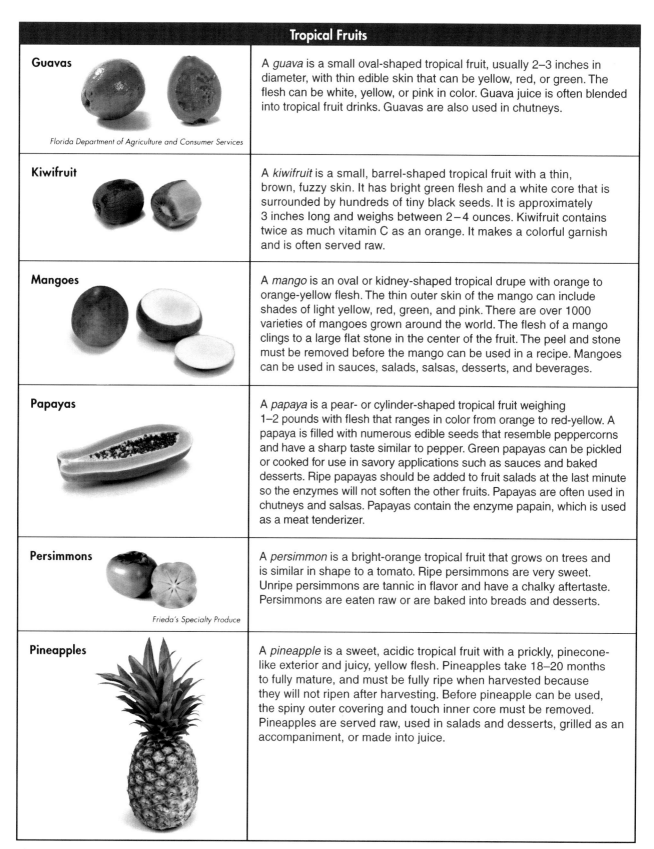

Tropical Fruits

Guavas

A *guava* is a small oval-shaped tropical fruit, usually 2–3 inches in diameter, with thin edible skin that can be yellow, red, or green. The flesh can be white, yellow, or pink in color. Guava juice is often blended into tropical fruit drinks. Guavas are also used in chutneys.

Florida Department of Agriculture and Consumer Services

Kiwifruit

A *kiwifruit* is a small, barrel-shaped tropical fruit with a thin, brown, fuzzy skin. It has bright green flesh and a white core that is surrounded by hundreds of tiny black seeds. It is approximately 3 inches long and weighs between 2–4 ounces. Kiwifruit contains twice as much vitamin C as an orange. It makes a colorful garnish and is often served raw.

Mangoes

A *mango* is an oval or kidney-shaped tropical drupe with orange to orange-yellow flesh. The thin outer skin of the mango can include shades of light yellow, red, green, and pink. There are over 1000 varieties of mangoes grown around the world. The flesh of a mango clings to a large flat stone in the center of the fruit. The peel and stone must be removed before the mango can be used in a recipe. Mangoes can be used in sauces, salads, salsas, desserts, and beverages.

Papayas

A *papaya* is a pear- or cylinder-shaped tropical fruit weighing 1–2 pounds with flesh that ranges in color from orange to red-yellow. A papaya is filled with numerous edible seeds that resemble peppercorns and have a sharp taste similar to pepper. Green papayas can be pickled or cooked for use in savory applications such as sauces and baked desserts. Ripe papayas should be added to fruit salads at the last minute so the enzymes will not soften the other fruits. Papayas are often used in chutneys and salsas. Papayas contain the enzyme papain, which is used as a meat tenderizer.

Persimmons

A *persimmon* is a bright-orange tropical fruit that grows on trees and is similar in shape to a tomato. Ripe persimmons are very sweet. Unripe persimmons are tannic in flavor and have a chalky aftertaste. Persimmons are eaten raw or are baked into breads and desserts.

Frieda's Specialty Produce

Pineapples

A *pineapple* is a sweet, acidic tropical fruit with a prickly, pinecone-like exterior and juicy, yellow flesh. Pineapples take 18–20 months to fully mature, and must be fully ripe when harvested because they will not ripen after harvesting. Before pineapple can be used, the spiny outer covering and touch inner core must be removed. Pineapples are served raw, used in salads and desserts, grilled as an accompaniment, or made into juice.

Figure 3-38. (continued on next page)

Tropical Fruits	
Plantains 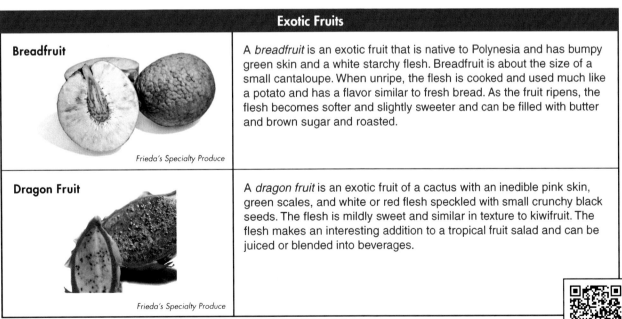	A *plantain* is a tropical fruit that is a close relative of the banana, but is larger and has a dark brown skin when ripening. When extremely ripe, the skin of a plantain turns black and the flesh is a soft, deep yellow. Unripe plantains are firm and starchy, like potatoes. Plantains are usually fried and are sometimes dried and ground into banana meal.
Pomegranates	The *pomegranate* is a round, bright-red tropical fruit with a hard, thick outer skin. Pomegranates measure about 3 inches in diameter and contain a thick white membrane that encloses hundreds of red seeds that are the edible portion of the pomegranate. The sweet and juicy seeds of the pomegranate are used in salads, soups, sauces, and beverages.
Prickly Pears *Frieda's Specialty Produce*	The *prickly pear* is a pear-shaped tropical fruit with protruding prickly fibers that is a member of the cactus family. The prickly pear is 2–4 inches long and has a very thick, coarse outer skin that can be green, yellow, orange, red, or deep purple, depending on the variety. The flesh is sweet and juicy and contains sweet, crisp, edible seeds. Peeled and sliced prickly pears are typically served cold.

Figure 3-38. Tropical fruits range in flavor from sweet to tangy and in texture from soft to crisp.

Exotic Fruits

An *exotic fruit* is type of fruit that comes from a hot, humid location but is not as readily available as a tropical fruit. Many exotic fruits are used in the professional kitchen. Breadfruit, dragon fruit, durians, jackfruit, kiwanos, kumquats, lychees, mangosteens, passion fruit, rambutans, and star fruit are considered exotic fruits due to their limited availability. **See Figure 3-39.**

Exotic Fruits	
Breadfruit *Frieda's Specialty Produce*	A *breadfruit* is an exotic fruit that is native to Polynesia and has bumpy green skin and a white starchy flesh. Breadfruit is about the size of a small cantaloupe. When unripe, the flesh is cooked and used much like a potato and has a flavor similar to fresh bread. As the fruit ripens, the flesh becomes softer and slightly sweeter and can be filled with butter and brown sugar and roasted.
Dragon Fruit *Frieda's Specialty Produce*	A *dragon fruit* is an exotic fruit of a cactus with an inedible pink skin, green scales, and white or red flesh speckled with small crunchy black seeds. The flesh is mildly sweet and similar in texture to kiwifruit. The flesh makes an interesting addition to a tropical fruit salad and can be juiced or blended into beverages.

Figure 3-39. (continued on next page)

Exotic Fruits

Exotic Fruits	
Durians *Frieda's Specialty Produce*	A *durian* is an exotic fruit that contains several pods of sweet, yellow flesh and has a custard-like texture. Durians are about the size of a large cantaloupe, have a thick thorny skin that is golden to green in color, and can weigh up to 10 pounds. This Malaysian fruit is known for having a strong, unpleasant odor but a very appealing taste.
Jackfruit *Florida Department of Citrus Melissa's Produce*	A *jackfruit* is an enormous, spiny, oval exotic fruit with yellow flesh that tastes like a banana and has seeds that can be boiled or roasted and then eaten. Jackfruit is the largest tree-borne fruit in the world. A jackfruit can weigh up to 100 pounds and be up to 36 inches long and approximately 20 inches in diameter. There are 100 –500 edible seeds in a single jackfruit. Immature jackfruit is similar in texture to chicken, making it a vegetarian substitute known as "vegetable meat." The flesh can be eaten raw or preserved in syrup. Like pineapple, jackfruit is a multiple fruit derived from the convergence of many individual flowers and a fleshy stem axis.
Kiwanos *Frieda's Specialty Produce*	A *kiwano,* also known as horned melon, is an exotic fruit with jagged peaks rising from an orange and red-ringed rind that is native to Africa. Its flesh is a brilliant lime-green color with cucumber-like seeds and a tart, refreshing kiwi-cucumber flavor. Kiwano melons can be cut in half lengthwise and eaten right out of the rind or sliced and added to fruit plates. The flesh can be strained to make a tangy-sweet sauce, salad dressing, or a refreshing beverage.
Kumquats *Frieda's Specialty Produce*	A *kumquat* is a small, golden, oval-shaped exotic fruit with a thin, sweet peel and tart center. A kumquat is similar in appearance to a small citrus fruit but is eaten peel and all. Kumquats are available from November to April. Kumquats are used to make jellies, marmalades, and chutneys. They are also an attractive garnish, especially when displayed with their leaves.
Lychees *United States Department of Agriculture*	A *lychee* is an exotic drupe covered with a thin, red, inedible shell and has a light-pink to white flesh that is refreshing, juicy, and sweet. The lychee also contains an inedible seed. Lychees must be harvested ripe, as they do not continue to ripen after being harvested. With the shell and seed removed, lychees can be eaten raw or in salads and desserts.

Figure 3-39. (continued on next page)

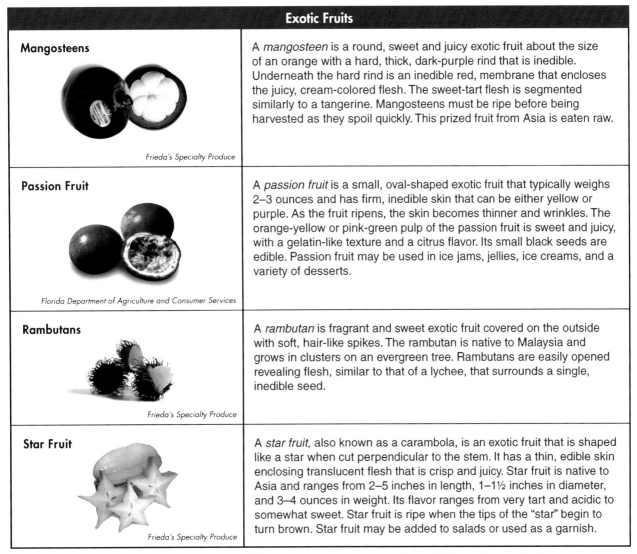

Exotic Fruits	
Mangosteens *Frieda's Specialty Produce*	A *mangosteen* is a round, sweet and juicy exotic fruit about the size of an orange with a hard, thick, dark-purple rind that is inedible. Underneath the hard rind is an inedible red, membrane that encloses the juicy, cream-colored flesh. The sweet-tart flesh is segmented similarly to a tangerine. Mangosteens must be ripe before being harvested as they spoil quickly. This prized fruit from Asia is eaten raw.
Passion Fruit *Florida Department of Agriculture and Consumer Services*	A *passion fruit* is a small, oval-shaped exotic fruit that typically weighs 2–3 ounces and has firm, inedible skin that can be either yellow or purple. As the fruit ripens, the skin becomes thinner and wrinkles. The orange-yellow or pink-green pulp of the passion fruit is sweet and juicy, with a gelatin-like texture and a citrus flavor. Its small black seeds are edible. Passion fruit may be used in ice jams, jellies, ice creams, and a variety of desserts.
Rambutans *Frieda's Specialty Produce*	A *rambutan* is fragrant and sweet exotic fruit covered on the outside with soft, hair-like spikes. The rambutan is native to Malaysia and grows in clusters on an evergreen tree. Rambutans are easily opened revealing flesh, similar to that of a lychee, that surrounds a single, inedible seed.
Star Fruit *Frieda's Specialty Produce*	A *star fruit,* also known as a carambola, is an exotic fruit that is shaped like a star when cut perpendicular to the stem. It has a thin, edible skin enclosing translucent flesh that is crisp and juicy. Star fruit is native to Asia and ranges from 2–5 inches in length, 1–1½ inches in diameter, and 3–4 ounces in weight. Its flavor ranges from very tart and acidic to somewhat sweet. Star fruit is ripe when the tips of the "star" begin to turn brown. Star fruit may be added to salads or used as a garnish.

Figure 3-39. Exotic fruits come from hot, humid locations but are not as readily available as tropical fruits.

Purchasing Fresh Fruit

Fresh fruit is packed in cartons, lugs, flats, crates, or bushels and is sold by weight or count. **See Figure 3-40.** Lugs can hold between 25–50 pounds of fruit, whereas flats are usually used to ship pint-size or quart-size containers of fruits such as blueberries or strawberries. Individual states have diverse regulations on the weight the containers in a flat can hold. Fruit size may also need to be specified when ordering from a vendor. For example, a 25 pound case of apples could contain 90, 110, or 125 apples, depending on the size of the fruit. Some fruits can also be purchased in a prepared form — cleaned, peeled, and cut.

Fresh Fruit

Figure 3-40. Fresh fruit is packed in cartons, lugs, flats, crates, or bushels and is sold by weight or count.

Ripening Fresh Fruit. It is important to know when fresh fruit should be purchased in order to purchase fruit with the best flavor and quality. As fruit begins to ripen, it changes in color and size and its flesh becomes soft and succulent. Left on the plant, fruit continues to ripen, breaking down in texture and flavor and eventually spoiling. Some fruits continue to ripen and mature after harvest, while other fruits do not. Fruits that continue to ripen after being harvested, such as bananas and pears, are often purchased before they are fully ripe. Other fruits, such as pineapples, have to be harvested fully ripe because they do not continue to ripen after they are picked.

The ripening process can be accelerated by storing fruit at room temperature or with other fruits that emit a large amount of ethylene gas. Apples, melons, and bananas give off ethylene gas and should be stored away from delicate fruits that could quickly ripen and spoil. Ripening can be delayed by chilling.

Fresh Fruit Grades. The USDA has a voluntary grading program for fresh fruit based on a variety of characteristics including uniformity of shape, size, color, texture, and the absence of defects. The USDA grades for fresh fruit include U.S. Fancy, U.S. No. 1, U.S. No. 2, and U.S. No. 3. Some fruit varieties have additional grades, such as U.S. Extra Fancy, U.S. Utility, or U.S. Commercial, that are specific to that particular variety. Most fruit used in restaurants is either U.S. Fancy or U.S. No. 1.

Some fruits are irradiated prior to being sold. *Irradiation* is the process of exposing food to low doses of gamma rays in order to destroy deadly organisms such as E. coli O157:H7, campylobacter, and salmonella. Irradiation also reduces spoilage bacteria, reduces insects and parasites such as fruit flies and the mango seed weevil, and in certain fruits and vegetables it inhibits sprouting and delays ripening. A statement indicating that the fruit was "treated with irradiation" and a radura, the international symbol of irradiation, appear on the label of irradiated foods. **See Figure 3-41.**

Radura Symbol

Figure 3-41. The radura symbol is required to be on the label of any food that has been irradiated.

Canned Fruit

Almost any type of fruit can be canned. Peaches, pears, and pineapples are commonly canned fruits. Fruit is canned at its peak. Depending on how long after harvest fresh fruit is eaten, canned fruit can have similar nutritional value. While the heating process does destroy some vitamins, canned fruits are adequate substitutes. Canned fruits that are packed in their own juice and those that have low amounts of added sugar and salt are the best choices.

Fruits are canned in water, light syrup, medium syrup, or heavy syrup. The packing method should be considered when using canned fruit in recipes, as the canning liquid can affect the flavor, texture, and nutritional value of the fruit.

Canned fruit can be stored for indefinite periods as long as it is kept in a cool, dry place. **See Figure 3-42.** Cans that are dented or bulging should be disposed of as they may contain harmful bacteria. After opening canned fruit, any unused portions should be transferred to an airtight storage container, dated and labeled, and then refrigerated.

Canned Fruit

InterMetro Industries Corporation

Figure 3-42. Canned fruits can be stored for indefinite periods when kept in a cool, dry place.

Frozen Fruit

Fruits, such as berries, are frozen whole, while other fruits, such as peaches, are cleaned, sliced, and frozen. **See Figure 3-43.** Freezing fruit inhibits the growth of microorganisms without affecting the nutritional value of the fruit. Individually quick-frozen fruits are convenient because the entire content need not be thawed to use a small portion. Some frozen fruits packaged in heavy syrup are also sold as purées for use in making ice creams, sorbets, and some sauces. The USDA grades for frozen fruit include U.S. Grade A or U.S. Fancy, U.S. Grade B or U.S. Choice, and U.S. Grade C or U.S. Standard.

Frozen Fruit

Figure 3-43. Frozen fruits, such as berries, are frozen whole and are convenient because the entire package need not be thawed in order to use a small portion.

Dried Fruit

Dried fruit is fruit that has had most of the moisture removed either naturally or through the use of a machine, such as a food dehydrator. Dried fruit has a much sweeter taste than fresh fruit due to the sugars being more concentrated. **See Figure 3-44.** Dried fruit is widely used in chutneys and compotes. It can also be added to breads and desserts. For maximum shelf life, dried fruit is stored in an airtight container in a cool, dry place.

Dried Fruit

Frieda's Specialty Produce

Figure 3-44. Dried fruits, such as cranberries, have a much sweeter taste than fresh fruit due to the sugars being more concentrated.

CHECKPOINT 3-3

1. Differentiate between variety fruits and hybrid fruits.

2. List the nutritional benefits of eating fruit.

3. Identify common berries used in the professional kitchen.

4. Identify three type of grapes used in the professional kitchen.

5. Identify common pomes used in the professional kitchen.

6. Identify common pears used in the professional kitchen.

7. Describe how quinces are typically cooked.

8. Identify common drupes (stone fruits) used in the professional kitchen.

9. Name the drupe that is harvested both for its fruit and its oil.

10. Identify common melons used in the professional kitchen.

11. Identify common citrus fruits used in the professional kitchen.

12. Identify common tropical fruits used in the professional kitchen.

13. Identify common exotic fruits used in the professional kitchen.

14. Describe factors to consider when purchasing fresh fruits.

15. Explain the role of canned fruits in the professional kitchen.

16. Explain the role of frozen fruits in the professional kitchen.

17. Explain the role of dried fruits in the professional kitchen.

COOKING FRUIT

Although fruit is often served raw, fruit can also be simmered, poached, grilled, broiled, baked, roasted, sautéed, and fried. Regardless of the cooking method used, it is important to remember that fruit is delicate and can become soft or mushy very quickly. Adding sugar or an acid such as lemon juice to fruit can help prevent it from becoming mushy in the cooking process. The sugar or acid is absorbed by the cells of the fruit, helping the fruit to plump up and stay firm. **See Figure 3-45.** However, adding an alkali such as baking soda will quickly break down the fruit, turning it to mush.

Cooking Fruit

Figure 3-45. Adding sugar to fruit helps the fruit stay plump.

Another characteristic of fruit to understand is the varying levels of pectin that different fruits contain. *Pectin* is a chemical present in all fruits that acts as a thickening agent when it is cooked in the presence of sugar and an acid. For example, when cranberries are simmered with sugar, they break down during the cooking process and pectin is released. As the mixture cools it thickens to a jelly-like consistency. Fruits that are high in pectin include apples, blackberries, cranberries, quinces, gooseberries, grapes, and plums. The peels of citrus fruits also contain a lot of pectin.

Pear Bureau Northwest

Simmering and Poaching Fruit

The simmering method is often used to make fruit compotes and stewed fruit. Fresh, frozen, canned, or dried fruit can be simmered. Simmering tenderizes and sweetens fruit. Simmered fruit can be served hot or cold and can accompany a dessert or entrée. **See Figure 3-46.** Apples, pears, peaches, and plums may be poached in water, liquor, wine, or syrup. Poaching is done at 185°F to ensure the fruit will retain its shape while being cooked.

Poached Fruit

Daniel NYC

Figure 3-46. Fruits such as pears may be poached in water, simple syrup, wine, or liquor.

Grilling and Broiling Fruit

Grilling or broiling fruit causes the sugars in the fruit to caramelize quickly because these two cooking methods require very high temperatures. Fruits that are good to grill or broil include pineapples, peaches, grapefruits, bananas, and pears. These fruits can be cut into slices or chunks and soaked in liquor or coated with sugar, honey, or liqueur for extra flavor before cooking. Fruit can be placed directly on the grill or cooked on skewers to make fruit kabobs. **See Figure 3-47.** When broiling fruit, the fruit should be placed on a sheet pan lined with parchment paper. Grilled or broiled fruit can be eaten alone or as an accompaniment.

Grilling Fruit

Chicken — Pineapple — Banana

National Chicken Council

Figure 3-47. Fruit can be added to skewers and grilled as kabobs.

Baking and Roasting Fruit

Most berries, pomes, and drupes are well suited for baking. Pies, tarts, cobblers, strudels, and turnovers are typically filled with fruits such as apples, blueberries, cherries, or peaches before they are baked. **See Figure 3-48.** The inner cavity of an apple, fig, or pear can also be stuffed with a flavorful filling. Fruit can also be added to meats that are being roasted. For example, ham is often covered with pineapple rings to add extra sweetness to the meat. Placing peach halves atop pieces of chicken during the final stages of roasting adds flavor and beauty to the dish.

Sautéing and Frying Fruit

Fruit is often sautéed in butter, sugar, spices, or liquor. The fruit develops a sweet, rich flavor and a syrupy, caramelized glaze. Sautéed fruit can be used in dessert dishes such as in crêpes or as toppings for ice cream. It can also be incorporated into savory mixtures that include garlic, onions, or shallots. Savory fruit mixtures pair well with entrées such as pork and poultry. **See Figure 3-49.**

Cooking Fruit

Baked Fruit Pies and Tarts

National Cherry Growers and Industries Foundation

Fruit Pie

Fruit Tart

Figure 3-48. Baked fruit pies and tarts are typically filled with fruits such as cherries, apples, blueberries, or peaches.

Nutrition Note

Baking with canned fruit produces similar nutritional values to baking with fresh fruit. Canned fruits may contain less vitamin C, but the canning process stabilizes other nutrients.

Sautéed Fruit

Courtesy of The National Pork Board

Figure 3-49. Sautéed fruits pair well with entrées such as pork.

Apples, bananas, pears, and peaches are suitable fruits for frying because they do not break down when exposed to very high temperatures. Before frying, the fruit is sliced into uniformly sized pieces so that it cooks evenly. The fruit is then patted dry with a paper towel to help the batter adhere to the fruit. Next, the fruit is dipped in the batter and fried in fat. When the batter turns golden brown, the fruit is removed from the hot fat and cooled on a rack while the excess fat drains off. Fried fruits may be garnished with confectioners' sugar or melted chocolate.

Plating Fruit

In some cases, fruit will be added to the plate instead of vegetables. In other cases, fruit may be plated with the protein, as in sautéed apples with roast pork. Fruit may also be served on a separate plate. **See Figure 3-50.** Fruit may be used in a complementary sauce, salsa, or chutney, such as a pineapple-mango salsa atop a sea bass fillet or a blueberry gastric served over lamb chops. As with all plating, attention should be paid to contrasting colors and textures. Cooked fruit should be caramelized for added color and flavor. Breakfast is an ideal time to add fruit to a plate.

Plated Fruit

National Honey Board

Figure 3-50. Fruit may be served on a separate plate or on the same plate as the entrée.

CHECKPOINT 3-4

1. Explain how the pectin level in a fruit affects the cooking process.

2. List common methods of cooking fruit.

3. Explain why fruit is poached at 185°F.

4. Simmer or poach a fruit and evaluate the quality of the prepared fruit.

5. Grill or broil a fruit and evaluate the quality of the prepared fruit.

6. Bake or roast a fruit and evaluate the quality of the prepared fruit.

7. Sauté or fry a fruit and evaluate the quality of the prepared fruit.

KNIVES

Knives are the most fundamental tool used in the professional kitchen. The use of a sharp knife in skilled hands can accomplish a wide variety of cutting tasks with great efficiency. Well-constructed knives are comfortable and balanced in the hand. Each part of a knife has a specific function. **See Figure 3-51.**

Knife Blades

Most knife blades used in the professional kitchen are made of high-carbon stainless steel or ceramic material. High-carbon stainless steel produces a blade that is easy to keep sharp, does not change color, and does not transfer a metallic taste to foods. Ceramic blades provide a sharper edge for a longer period than any other material, do not react with acidic foods, and are very easy to keep clean. Ceramic blades are made from zirconium oxide, a less flexible material than stainless steel, and may chip or break if they strike hard surfaces, such as large bones, or are dropped on tile floors.

Kyocera Advanced Ceramics

Parts of a Knife

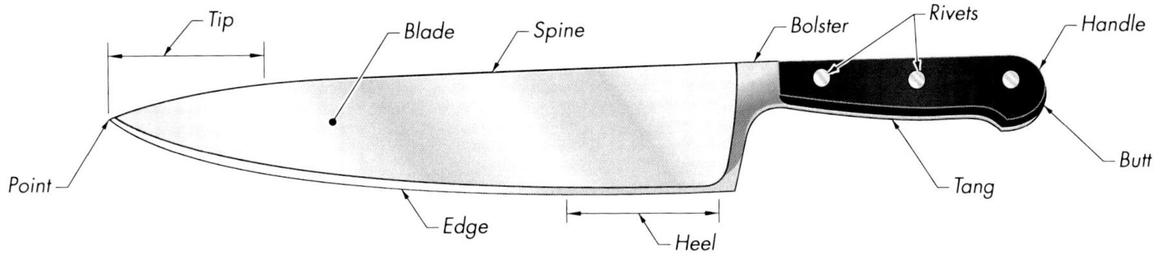

Figure 3-51. Each part of a knife, including all parts of the blade, as well as the tang, handle, bolster, and rivets, has a specific function.

A knife blade consists of five parts: the heel, tip, point, spine, and edge.

- The *heel* is the rear portion of the knife blade and is most often used to cut thick items where more force is required.
- The *tip* is the front quarter of the knife blade. Most cutting is accomplished with the section of the blade between the tip and the heel.
- The point of the blade is used as a piercing tool.

- The *spine* is the unsharpened top part of the knife blade that is opposite the edge.
- The *edge* is the sharpened part of the knife blade that extends from the heel to the tip. A knife with a sharp edge is safer than a knife with a dull edge because it requires less pressure to use. There are four basic types of knife blade edges: straight, serrated, granton, and hollow ground. **See Figure 3-52.**

Knife Blade Edges	
Straight 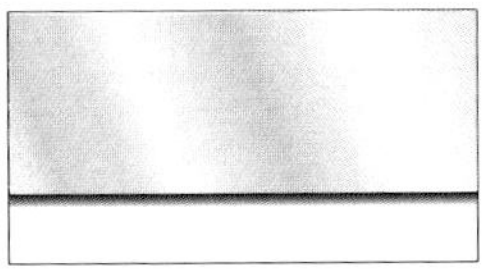	Straight edge blades are the most common type of knife blade.
Serrated 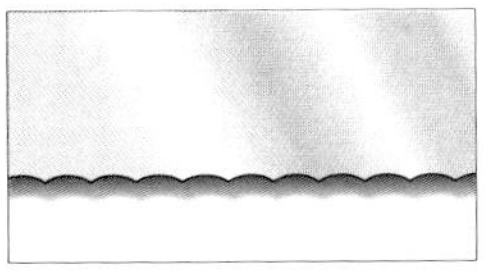	Serrated edge blades have scallop-shaped teeth that easily penetrate tough outer crusts or skins of food products such as breads and fruits.
Granton 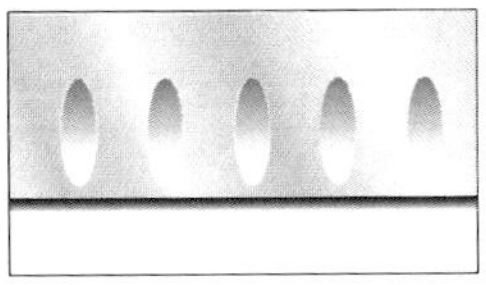	Granton edge blades have hollowed out grooves running along both sides that reduce the amount of friction as the edge of the blade cuts the food, allowing maximum contact. Granton edge blades are often used to cut meats and poultry.
Hollow Ground 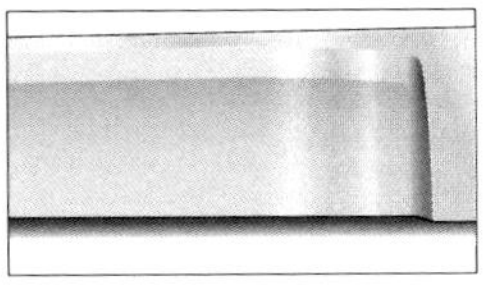	Hollow ground edge blades have been ground just below the midpoint of the blade to form a very thin cutting edge that is easily dulled. Hollow ground edge blades are ideal for skinning fish, peeling fruits, and preparing sushi.

Figure 3-52. Each of the four basic types of knife blade edges offers an advantage when cutting specific foods.

Tangs

The *tang* is the unsharpened tail of a knife blade that extends into the handle. The highest-quality knives have a tang that extends all the way to the end of the handle. The tang contains holes for securing the handle to the blade. A *partial tang* is a shorter tail of a knife blade that has fewer rivets than a full tang. Partial tang knives are less durable than full-tang knives, but may be acceptable for infrequent or light use. A *rat-tail tang* is a narrow rod of metal that runs the length of the knife handle but is not as wide as the handle. Rat-tail tangs are fully enclosed in the handle and are less durable than full or partial tangs.

Knife Handles

The handle of a knife can be made from wood, stainless steel, or synthetic materials such as plastic, nylon, styrene, resin, or polypropylene. Wood handles are becoming less common due to their lack of durability and how easily they trap bacteria. Stainless steel handles are virtually maintenance free, however, they become slippery when wet. Synthetic handles are easy to clean, last longer than wood, and are easier to grip when wet than stainless steel. However, synthetic materials crack over time and when they are exposed to extreme temperature changes. The end of a knife handle is referred to as the butt of the knife.

Production Tip

Knives should always be washed by hand. Knives should never be left in standing water or placed in a commercial dishwasher because this can cause the handles to crack or warp.

Bolsters and Rivets

A *bolster* is a thick band of metal located where the blade of a knife joins the handle. The purpose of the bolster is to provide strength to the blade and prevent food from entering the seam between the blade and the handle. A *rivet* is a metal fastener used to attach the tang of a knife to the handle. Some knives do not have bolsters and rivets.

Large Knives

Many different types of knives are used in the professional kitchen. Knowing which knife to use in a given application makes working with knives safer and more efficient. Large knives are used to make precise cuts in large areas or to separate food items such as meats. Large knives used in the professional kitchen include boning knives, bread knives, butcher's knives, chef's knives, cleavers, santoku knives, scimitars, slicers, and utility knives. **See Figure 3-53.**

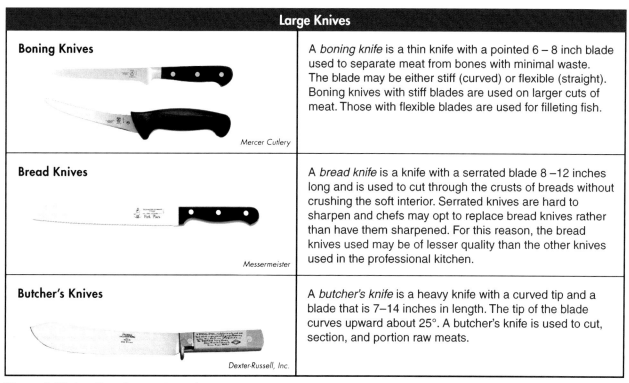

Large Knives	
Boning Knives *Mercer Cutlery*	A *boning knife* is a thin knife with a pointed 6 – 8 inch blade used to separate meat from bones with minimal waste. The blade may be either stiff (curved) or flexible (straight). Boning knives with stiff blades are used on larger cuts of meat. Those with flexible blades are used for filleting fish.
Bread Knives *Messermeister*	A *bread knife* is a knife with a serrated blade 8 –12 inches long and is used to cut through the crusts of breads without crushing the soft interior. Serrated knives are hard to sharpen and chefs may opt to replace bread knives rather than have them sharpened. For this reason, the bread knives used may be of lesser quality than the other knives used in the professional kitchen.
Butcher's Knives *Dexter-Russell, Inc.*	A *butcher's knife* is a heavy knife with a curved tip and a blade that is 7–14 inches in length. The tip of the blade curves upward about 25°. A butcher's knife is used to cut, section, and portion raw meats.

Figure 3-53. (continued on next page)

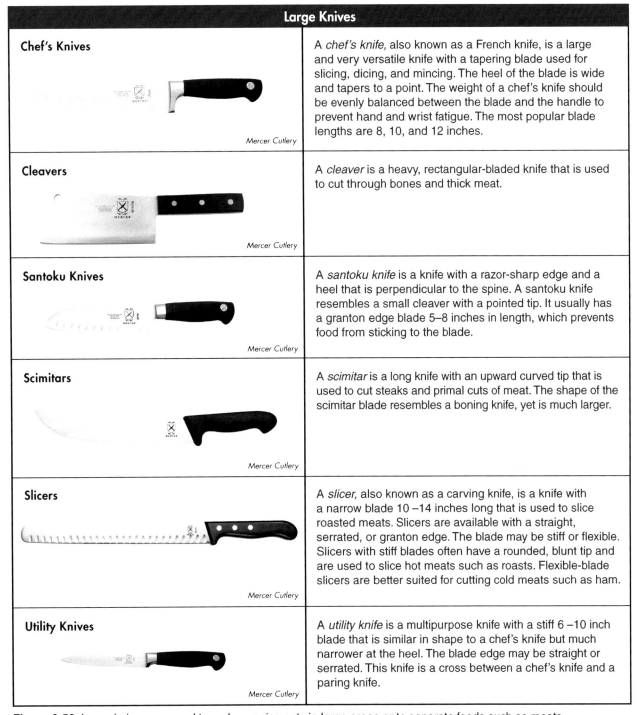

Large Knives	
Chef's Knives *Mercer Cutlery*	A *chef's knife,* also known as a French knife, is a large and very versatile knife with a tapering blade used for slicing, dicing, and mincing. The heel of the blade is wide and tapers to a point. The weight of a chef's knife should be evenly balanced between the blade and the handle to prevent hand and wrist fatigue. The most popular blade lengths are 8, 10, and 12 inches.
Cleavers *Mercer Cutlery*	A *cleaver* is a heavy, rectangular-bladed knife that is used to cut through bones and thick meat.
Santoku Knives *Mercer Cutlery*	A *santoku knife* is a knife with a razor-sharp edge and a heel that is perpendicular to the spine. A santoku knife resembles a small cleaver with a pointed tip. It usually has a granton edge blade 5–8 inches in length, which prevents food from sticking to the blade.
Scimitars *Mercer Cutlery*	A *scimitar* is a long knife with an upward curved tip that is used to cut steaks and primal cuts of meat. The shape of the scimitar blade resembles a boning knife, yet is much larger.
Slicers *Mercer Cutlery*	A *slicer,* also known as a carving knife, is a knife with a narrow blade 10 –14 inches long that is used to slice roasted meats. Slicers are available with a straight, serrated, or granton edge. The blade may be stiff or flexible. Slicers with stiff blades often have a rounded, blunt tip and are used to slice hot meats such as roasts. Flexible-blade slicers are better suited for cutting cold meats such as ham.
Utility Knives *Mercer Cutlery*	A *utility knife* is a multipurpose knife with a stiff 6 –10 inch blade that is similar in shape to a chef's knife but much narrower at the heel. The blade edge may be straight or serrated. This knife is a cross between a chef's knife and a paring knife.

Figure 3-53. Large knives are used to make precise cuts in large areas or to separate foods such as meats.

Small Knives

Small knives are used to make precise cuts in small areas or to open food items such as shellfish. Small knives commonly used in the professional kitchen include clam knives, oyster knives, paring knives, and tourné knives. **See Figure 3-54.**

Special Cutting Tools

In addition to knives, special cutting tools are used to cut food items for specific applications. Although there are many special cutting tools, those commonly used in the professional kitchen include channel knives, mandolines, parisienne scoops, peelers, and zesters. **See Figure 3-55.**

Small Knives	
Clam Knives 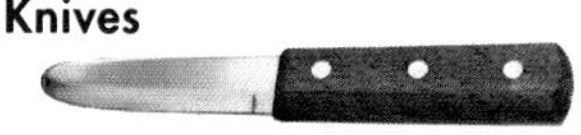 *American Metalcraft, Inc.*	A *clam knife* is a small knife with a short, flat, round-tipped sharp blade that is used to open clams. The proper use of a clam knife makes the task of opening clams an efficient and safe process.
Oyster Knives  *Browne-Halco (NJ)*	An *oyster knife* is a small knife with a short, dull-edged blade with a tapered point that is used to open oysters. The proper use of an oyster knife makes the task of opening oysters an efficient and safe process.
Paring Knives *Mercer Cutlery*	A *paring knife* is a short knife with a stiff 2– 4 inch blade used to trim and peel fruits and vegetables. A paring knife is often used in conjunction with a chef's knife to remove stems from produce.
Tourné Knives  *Mercer Cutlery*	A *tourné knife,* also known as a bird's beak knife, is a short knife with a curved blade that is primarily used to carve vegetables into a specific shape called a tourné, which is a seven-sided football shape with flat ends.

Figure 3-54. Small knives are used to make precise cuts in small areas or to open food items such as shellfish.

Special Cutting Tools	
Channel Knives  *Dexter-Russell, Inc.*	A *channel knife* is a special cutting tool with a thin metal blade within a raised channel that is used to remove a large string from the surface of a food item. A channel knife leaves a decorative pattern on the surface of an item, such as a cucumber.
Mandolines 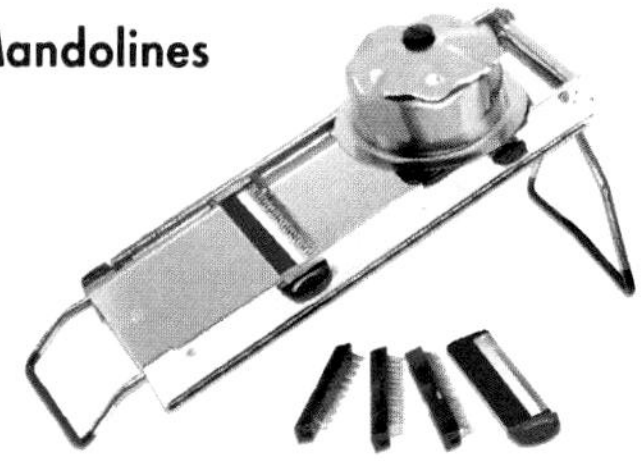 *Paderno World Cuisine*	A *mandoline* is a special cutting tool with adjustable steel blades used to cut food into consistently thin slices. A mandoline can cut foods paper thin and also produce julienne cuts and waffle cuts. A hand guard needs to be in place when using a mandoline.
Parisienne Scoops  *Paderno World Cuisine*	A *parisienne scoop* is a special cutting tool that has a half-ball cup with a blade edge attached to a handle and is used to cut fruits and vegetables into uniform spheres.
Peelers  *Carlisle FoodService Products*	A *peeler* is a special cutting tool with a swiveling, double-edged blade that is attached to a handle and is used to remove the skin or peel from fruits and vegetables. The double-edged blade contours to the shape of the fruit or vegetable.
Zesters  *Browne-Halco (NJ)*	A *zester* is a special cutting tool with tiny blades inside of five or six sharpened holes that are attached to a handle. To use a zester, the cutting holes are drawn across the peel of a citrus fruit to yield small strings or "zest" that can be added to foods as a natural flavoring.

Figure 3-55. Special cutting tools are used to cut food items for specific applications.

Knife Safety

Improper use of knives can lead to injury. The following safety precautions should be taken when holding, using, carrying, washing, and storing knives:

- Always grip a knife properly to ensure safety and control. When using a knife, the more pressure that is applied the higher the risk of the knife slipping and of personal injury occurring.
- Always position the guiding hand properly when using, sharpening, and honing a knife.
- Always cut food items on a nonporous cutting board because the nonporous surface greatly reduces the risk of cross-contamination. Color-coded cutting boards may be used for specific types of foods. **See Figure 3-56.**
- Always pass a knife to a person by laying it on a table and sliding it forward.
- When walking with a knife, always keep the knife pointing down and hold it along the side of the body.
- Use only clean, sanitized knives on a whetstone or sharpening steel to avoid cross-contamination.
- Always wipe a blade after using a whetstone or sharpening steel to remove any metal residue.
- Always keep knives sharp. Injury is more likely to occur with a dull knife than a sharp one. Hone knives after each use to maintain a smooth, sharp edge.
- Always wipe a knife blade with the edge facing away from the hand.
- Always clean and sanitize knives before storing them.
- Store knives in sleeves, guards, or knife holders to avoid injury.
- Never leave knives in a sink as someone could reach in and be injured.
- Never wash knives in a commercial dish machine because the heat and chemicals can ruin the handles.
- Never use a knife to pry a lid off of any type of container.
- Never attempt to catch a falling knife.

Knife Grip and Positioning

There are different acceptable methods for gripping a knife, but there is a common method used by culinary professionals that provides control and stability. To begin, the knife is held by the handle while resting the side of the index finger against one side of the blade and placing the thumb on the other side of the blade. The hand not holding the knife is referred to as the guiding hand. The guiding hand is responsible for guiding the item to be cut into the knife. To correctly position the fingers of the guiding hand, imitate the shape of a spider on the table. The fingertips should all be slightly tucked, yet touch the surface of the table. This guiding hand position is used to safely hold the food next to the blade of the knife.

Cutting Boards

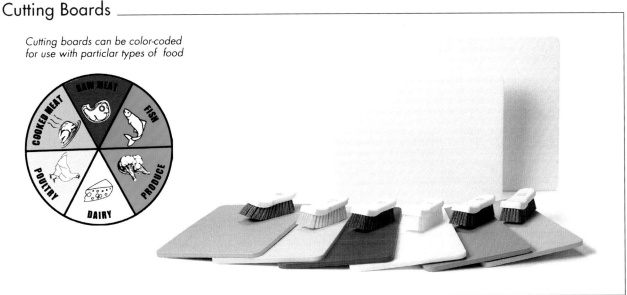

Cutting boards can be color-coded for use with particlar types of food

Carlisle FoodService Products

Figure 3-56. Color-coded, nonporous cutting boards may be used for specific types of foods to reduce the risk of cross-contamination.

Using the proper knife grip, the tip of the knife is placed on the cutting board. The guiding hand is placed next to the knife blade in the proper position, with fingertips slightly tucked under near the back half of the blade. The side of the blade should rest against the knuckle of the middle finger of the guiding hand. This position reduces the chances of cutting fingers. **See Figure 3-57.**

Knife Grip and Positioning

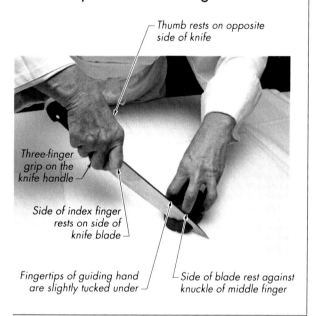

Figure 3-57 labels:
- Thumb rests on opposite side of knife
- Three-finger grip on the knife handle
- Side of index finger rests on side of knife blade
- Fingertips of guiding hand are slightly tucked under
- Side of blade rest against knuckle of middle finger

Figure 3-57. Using the proper knife grip with the knife hand, and with fingertips slightly tucked under with the guiding hand, the side of the blade should rest against the knuckle of the middle finger of the guiding hand.

With the proper knife grip and hand position, a rocking motion is used to cut with a chef's knife. Using the wrist as a fulcrum, the handle is brought down as the tip of the knife slides forward. Likewise, the handle is raised up as the tip of the knife slides backward. This rocking movement, coupled with the correct position of the guiding hand, creates a controlled motion that can be used to efficiently cut through food.

Production Tip

When cutting using a rocking motion, continually rest the blade of the knife against the knuckle of the middle finger of the guiding hand.

Procedure for Using Proper Cutting Techniques

1. Using the proper cutting grip, place the tip of the knife on the cutting board and press down on the knife handle.

2. Continue pressing down on the handle and slide the blade forward, following the curve of the blade.

3. With the heel of the blade on the cutting surface, slide the blade backward and then raise the handle slightly to position the knife for the next slice.

Sharpening Knives

Always check the edge of a knife to make sure it is sharp and properly maintained before using it. A sharp knife is much safer than a dull knife. Having a sharp knife helps to prevent injury because less pressure is required to use a sharp knife as compared to a dull knife.

Although hand-held or electric sharpeners can be used, a whetstone is typically used to sharpen professional knives. A *whetstone* is a stone used to grind the edge of a blade to the proper angle for sharpness. A three-sided whetstone has a coarse-grit, a medium-grit, and a fine-grit side. Two-sided whetstones have a medium-grit side and a fine-grit side.

To sharpen a knife, the blade of the knife is held at a specific angle to the stone. To achieve this angle, the knife blade is held at a 90° angle straight above the whetstone as if it were cutting the stone in half. Then, the knife is tilted halfway toward the stone, at a 45° angle, and then halfway again to find the perfect sharpening angle between 20° and 25°. **See Figure 3-58.**

Sharpening Angle

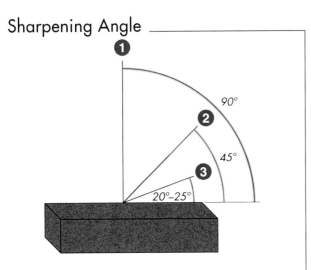

1 Hold the knife blade straight above the whetstone as if cutting the stone in half.

2 Tilt the blade halfway toward the stone to reach a 45° angle.

3 Tilt the blade halfway again to the correct sharpening angle of 20–25°.

Figure 3-58. To sharpen a knife, the blade is held at a 20–25° angle against the whetstone.

After the proper angle is achieved, the knife blade is then slowly dragged across the stone from tip to heel while applying light pressure. **See Figure 3-59.**

Sharpening Knives

1 Lay the edge of the knife near the top corner of the whetstone at a 20–25° angle.

2 Starting at the knife tip, slowly draw the blade across the surface of the stone at a 20–25° angle until reaching the heel.

3 Flip the knife over and repeat process on the other side, using the same number of strokes to create an even and sharp edge.

Figure 3-59. After establishing the correct angle, the blade is dragged across the whetstone from tip to heel while applying light pressure.

Honing Knives

After using a whetstone, it is important to "hone" or align the edge of a knife blade. *Honing,* also known as truing, is the process of aligning a blade's edge and removing any burrs or rough spots on the blade. A *steel,* also known as a butcher's steel, is a steel rod approximately 18 inches long attached to a handle and is used to align the edge of knife blades.

Sharpening Knives
Media Clip

The 20–25° angle used to sharpen knives is also used to hone knives. To achieve this angle, hold the steel perpendicular or pointed toward the floor with the guiding hand and hold the knife blade at a 20–25° angle in relation to the steel. This can be done by first holding the blade at a 90° angle to the steel, then adjusting it to about half that angle (45°), and finally adjusting it about half of that angle again to a 20–25° angle. **See Figure 3-60.** A steel should always be used to hone the blade of a knife after sharpening as well as between sharpenings to maintain a sharp, smooth edge.

Positioning Knife for Honing

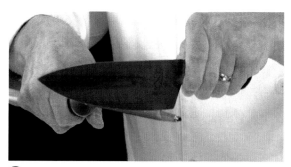

1 *Hold the blade at a 90° angle to the steel.*

2 *Adjust the blade to half that angle (45°).*

3 *Adjust the blade about halfway again to reach the correct 20–25° angle.*

Figure 3-60. To hone a knife, the blade is held at a 20–25° angle against the steel.

CHECKPOINT 3-5

1. Identify the main parts of a knife and explain the function of each.

2. Explain why a knife with a sharp edge is safer to use than a knife with a dull edge.

3. Explain the advantage of using a knife with a granton edge blade.

4. Contrast high-carbon stainless steel blades and ceramic blades.

5. Differentiate between a partial tang and a rat-tail tang.

6. Describe eight types of large knives used in the professional kitchen.

7. Describe four types of small knives used in the professional kitchen.

8. Describe five special cutting tools used in the professional kitchen.

9. Explain why nonporous cutting boards are used in the professional kitchen.

10. Describe how to safely carry a knife when walking.

11. Describe how to safely pass a knife to another person.

12. Describe how knives should be stored.

13. Provide four examples of what should never be done with a knife.

14. Demonstrate how to properly grip and position a knife.

15. Identify the term used to describe the hand that is not used to hold the knife.

16. Identify the angle at which a knife blade is held against a whetstone when being sharpened.

17. Explain the process of honing a knife.

18. Sharpen and hone a chef's knife.

BASIC KNIFE CUTS

Every foodservice professional must know the dimensions of the basic knife cuts and be able to execute them accurately and efficiently. Uniform knife cuts ensure that items cook evenly and look appealing in the finished product. Common knife cuts used in the professional kitchen can be grouped into slicing cuts, stick cuts, dice cuts, mincing, chopping, fluted cuts, and tourné cuts.

Slicing Cuts

Slicing involves passing the blade of the knife slowly through an item to make long, thin pieces. In slicing, the knife is pulled backward or slid forward through the item. Slicing cuts include the rondelle, diagonal, oblique, and chiffonade.

Rondelle Cuts. A *rondelle cut,* also known as a round cut, is a slicing cut that produces disks. Rondelle cuts are produced from slicing cylindrical vegetables such as cucumbers and carrots straight through. To make a rondelle cut, the cylindrical vegetable is placed perpendicular to blade and then sliced to create ¼, ⅛, or 1⁄16 inch disks. **See Figure 3-61.**

Rondelle Cuts

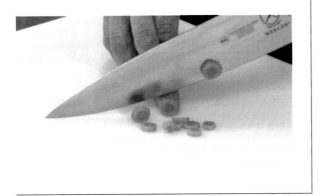

Figure 3-61. A rondelle cut, also known as a round cut, is a slicing cut that produces disks.

Diagonal Cuts. A *diagonal cut* is a slicing cut that produces flat-sided, oval slices. Diagonal cuts are made from cylindrical vegetables that are cut on the bias. To make a diagonal cut, place the item at a 45° angle to the knife blade. Then, guide the item toward the blade as each cut is made. **See Figure 3-62.**

Diagonal Cuts

Figure 3-62. A diagonal cut is a slicing cut that produces flat-sided, oval slices.

Oblique Cuts. An *oblique cut,* also known as a rolled cut, is a slicing cut that produces wedge-shaped pieces with two angled sides. The sides are neither parallel nor perpendicular. The oblique cut is similar to the diagonal cut in that 45° angle slices are also made on a cylindrical item. However, the oblique cut produces larger pieces that are more wedge-shaped.

Chiffonade Cuts. A *chiffonade cut* is a slicing cut that produces thin shreds of leafy greens or herbs. Chiffonade-cut items can be used as ingredients or as a base under displayed foods. To make a chiffonade cut, first wash the leafy items, stack the leaves on top of one another, and roll the stack lengthwise like a cigar. Place the cigar-shaped roll on the cutting board perpendicular to the knife blade. Use a rocking motion to thinly slice the roll as it is fed with the guiding hand into the knife blade. The result is finely shredded leaves or herbs.

Procedure for Making Oblique Cuts

1. With fingers of the guiding hand tucked, guide the cylindrical vegetable toward the blade at a 45° angle and slice. Reserve this first cut for later use.
2. Roll the vegetable 180° and slice again, keeping the blade at a 45° angle. *Note:* This slice produces the first oblique cut.

3. Roll the item back 180° to the original position and slice again.

4. Continue rolling the item 180° between cuts to make wedge-shaped oblique cuts.

Procedure for Making Chiffonade Cuts

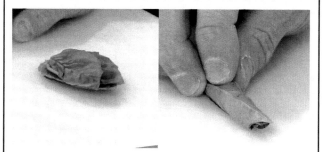

1. Place washed, dry leaves in a neat stack.
2. Roll the stack into a tight cylinder and place the cylinder perpendicular to the knife blade.

3. With the fingers of the guiding hand tucked, use a rocking motion to thinly slice the leaves.

Finished basil chiffonades are very finely cut.

Stick Cuts

Many fruits and vegetables are cut into sticklike shapes to create a uniform appearance and to ensure even cooking. Common stick cuts include batonnet, julienne, and fine julienne cuts.

- A *batonnet cut* is a stick cut that produces a stick-shaped item ¼ × ¼ × 2 inches long.
- A *julienne cut* is a stick cut that produces a stick-shaped item ⅛ × ⅛ × 2 inches long.
- A *fine julienne cut* is a stick cut that produces a stick-shaped item 1/16 × 1/16 × 2 inches long.

Oblique Cuts Media Clip

Chiffonade Cuts Media Clip

Julienne Cuts Media Clip

Procedure for Making Batonnet Cuts

1. Cut washed and peeled vegetables, such as carrots, into pieces 2 inches in length.
2. Carefully square off three sides, leaving the fourth side rounded. Save scraps for later use.

4. Stack a few ¼ inch planks.
5. Carefully slice again into ¼ inch sticks.

3. With the rounded side farthest away from the knife, cut even, ¼ inch thick planks. Save scraps.

Finished batonnets measure ¼ × ¼ × 2 inches.

All stick cuts begin by squaring off the item to be cut. There are industry-accepted dimensions for each stick cut. However, the length of stick cuts may vary depending on the desired result for a specific dish.

Dice Cuts

Dice cuts are precise cubes cut from uniform stick cuts. To produce a dice cut, a stick cut of the appropriate dimension is cut into cubes. **See Figure 3-63.** Common dice cuts include large dice, medium dice, small dice, brunoise, and fine brunoise. Items, such as onions, consist of many layers and require a modified procedure for dicing.

- A large dice is ¾ × ¾ × ¾ inch cubes cut from ¾ × ¾ × 2 inch sticks.
- A medium dice is ½ × ½ × ½ inch cubes cut from ½ × ½ × 2 inch sticks.
- A small dice is ¼ × ¼ × ¼ inch cubes cut from ¼ × ¼ × 2 inch sticks, or batonnets.

Dice Cuts

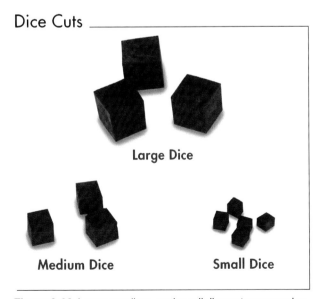

Figure 3-63. Large, medium, and small dice cuts are precise cubes cut from uniform stick cuts.

Batonnet Cuts
Media Clip

Procedure for Dicing Onions

1. Using a chef's knife, cut off the stem end and lightly trim the root end of an onion. *Note:* Do not cut the root end off completely as it holds the layers of the onion in place, preventing it from falling apart.

2. Cut the onion in half from the stem end to root end.

3. Make a thin slice from root end to stem end through the outer peel only.
4. Use the tip of the paring knife to pull off the top layer of the peel.

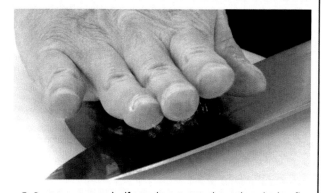

5. Position onion half on the cutting board with the flat side down. Use the chef's knife to make two or three horizontal cuts through the onion, leaving the root end intact.

6. Make vertical slices through the onion from stem end to root end, again leaving the root end intact. *Note:* The closer together the slices, the smaller the finished dice.

7. Turn the onion a quarter turn and make cuts the thickness of the desired dice, slicing all the way through from stem end to root end. Repeat the dicing process on the other half of the onion.

Dicing Onions
Media Clip

A *brunoise cut* is a dice cut that produces a cube-shaped item with six equal sides measuring ⅛ inch each. A brunoise is cut from a julienne stick. A *fine brunoise cut* is a dice cut that produces a cube-shaped item with six equal sides measuring ¹⁄₁₆ inch. A fine brunoise is cut from a fine julienne stick. **See Figure 3-64.**

Procedure for Making Brunoise Cuts

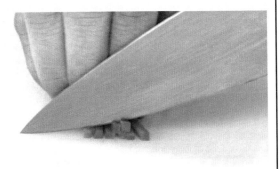

1. Place a small bundle of julienne sticks perpendicular to the knife blade.

2. With a slicing motion, cut through the bundle at equally spaced intervals to produce six-sided cubes.

Finished brunoise cuts measure ⅛ × ⅛ × ⅛ inches.

Fine Brunoise Cuts

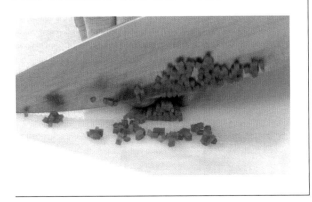

Figure 3-64. A fine brunoise cut is a dice cut that uses a fine julienne stick to produce a cube-shaped item with six equal sides measuring ¹⁄₁₆ inch each.

Chopping

Chopping is rough-cutting an item so that there are relatively small pieces throughout, although there is no uniformity in shape or size. Parsley, hard-cooked eggs, and a rough-cut mix of vegetables called mirepoix are often chopped because a uniform shape is not required.

Mincing

Mincing and chopping have fewer applications in the professional kitchen than the other knife cuts. Mincing is finely chopping an item to yield a very small cut, yet not entirely uniform, product. Shallots, garlic, and fresh herbs are commonly minced.

Brunoise Cuts
Media Clip

Procedure for Mincing

1. Using a paring knife, cut off the stem end and lightly trim the root end of a vegetable, such as a shallot. *Note:* Do not cut the root end off completely as it holds the layers of the shallot in place, preventing it from falling apart.

2. Make a thin slice from root end to stem end through the outer peel only.

3. Use the tip of the paring knife to pull off the top layer of the peel.

4. Cut the shallot in half lengthwise from stem end to root end.

5. Lay half of the shallot on the cutting board with the flat side down. Use the tip of the knife to make two or three horizontal cuts through the shallot, leaving the root end intact.

6. Make vertical slices through the shallot from stem end to root end, again leaving the root end intact. *Note:* The closer together the slices, the smaller the finished dice.

7. Using a chef's knife, turn the shallot and make cuts all the way through from stem end to root end until only the root remains.

8. Place the guiding hand, opened and flat, on the top of the blade to help pivot the knife back and forth.

Mincing
Media Clip

Fluted Cuts

A *fluted cut* is a specialty cut that leaves a spiral pattern on the surface of an item by removing only a sliver with each cut. Button mushrooms are often fluted. Making fluted cuts requires good hand-eye coordination as each sliver is cut away and then the mushroom is turned slightly clockwise or counter clockwise before making the next cut from the same central point at the top of the mushroom.

Tourné Cuts

Tourné is a French word from the verb "to turn." A *tourné cut* is a carved, football-shaped cut with seven sides and flat ends. Turnips, carrots, and beets are often displayed as tournés on elegant plate presentations. This challenging specialty cut requires a lot of practice to master. Tournés are viewed as a demonstration of masterful knife skills.

Procedure for Cutting Flutes

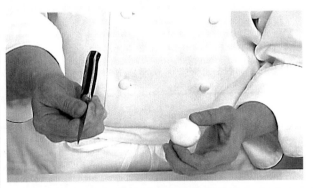

1. Wipe mushroom with a damp towel. *Note:* Always start with a cold button mushroom.
2. Hold the mushroom with the index finger and thumb.
3. While holding a paring knife, place the thumb on the top of the blade, the index finger on the bottom of the blade, and the middle finger underneath the knife to hold it steady.

4. Hold the mushroom level and place the front third of knife blade on the center point of the top of the mushroom facing 12 o'clock.

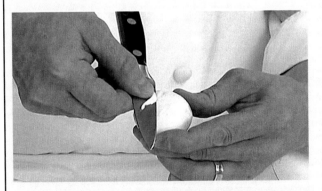

5. Turn the knife to 7 o'clock (right-handed) or 5 o'clock (left-handed), keeping the knife in contact with the mushroom at all times. *Note:* A thin strip of mushroom will be removed with each cut, leaving a fluted edge.

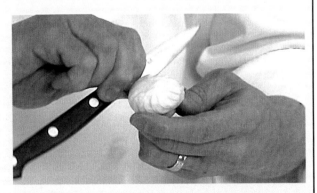

6. Rotate the mushroom clockwise (right-handed) or counter clockwise (left-handed) and continue cutting flutes until they intersect at the top center of the mushroom.

Flutes
Media Clip

Procedure for Cutting Tournés

1. Cut a washed and peeled root vegetable, such as a beet or a carrot, into 2 inch long pieces. *Note:* Wider vegetables, such as potatoes, may need to be cut into 4–6 sections before being cut into 2 inch long pieces.
2. Cut each 2 inch piece into 1 inch widths.

5. Turn the vegetable slightly and continue to carve the item in smooth, curved strokes until a seven-sided football shape is achieved. *Note:* As each slice is carved, the ends will become narrower than the middle.

3. Holding the item in the guiding hand, place the index finger on one end and the thumb on the opposite end.
4. Using a tourné knife, slowly carve the item from one end to the other in a smooth, continuous stroke that creates a slightly rounded surface.

A finished tourné has seven sides and is flat on each end.

CHECKPOINT 3-6

1. Explain why uniform knife cuts are used in the professional kitchen.

2. Describe the difference between a diagonal cut and an oblique cut.

3. Slice a root vegetable into rondelles, diagonals, and obliques.

4. Chiffonade a bunch of fresh greens.

5. Cut a root vegetable into batonnet, julienne, and fine julienne stick cuts.

6. Cut a whole fruit into large dice, medium dice, and small dice.

7. Dice an onion.

8. Cut julienne sticks into brunoise cubes and fine julienne sticks into fine brunoise cubes.

9. Differentiate between mincing and chopping.

10. Mince a shallot.

11. Flute a button mushroom.

12. Tourné a turnip, a beet, or a carrot.

Tournés
Media Clip

POTATO CLASSIFICATIONS

Potatoes are often referred to as starches in the professional kitchen. A *potato* is a round, oval, or elongated tuber that is the only edible part of the potato plant. The color of potato skin differs among varieties and can be brown, white, purple, red, or yellow. Depending on the variety, potato flesh can be creamy white, yellow-gold, or purple in color. A potato eaten with the skin on provides dietary fiber, vitamin C, vitamin B6, and potassium. Mealy potatoes, waxy potatoes, and new potatoes, as well as sweet potatoes and yams are used in the professional kitchen.

Mealy Potatoes

A *mealy potato* is a type of potato that is higher in starch and lower in moisture than other types of potatoes. Mealy potatoes are the preferred type of potato for baking, frying, mashing, puréeing, and making casseroles. After cooking, these potatoes become light and fluffy inside. Mealy potatoes include purple potatoes, russet potatoes, and white potatoes. **See Figure 3-65.**

Waxy Potatoes

A *waxy potato* is a type of potato with a thin skin and slightly waxy flesh that is lower in starch and higher in moisture than mealy potatoes. Compared to mealy potatoes, waxy potatoes stay much firmer in the center when fully cooked and also retain their shape better. Waxy potatoes can be roasted, sautéed, steamed, or simmered. Waxy potatoes are the common choice for cooking except when baking or deep-frying. Fingerling potatoes, red potatoes, and yellow potatoes are waxy potatoes. **See Figure 3-66.**

New Potatoes

A *new potato,* also known as an early crop potato, refers to any variety of potato that is harvested before the sugar is converted to starch. Because they are harvested so early, new potatoes are small and relatively uniform in size. **See Figure 3-67.** They have a thin, delicate skin and tender flesh. New potatoes hold their shape after cooking. They are often roasted, steamed, or simmered. Due to their high moisture content, new potatoes spoil more quickly than other types of potatoes.

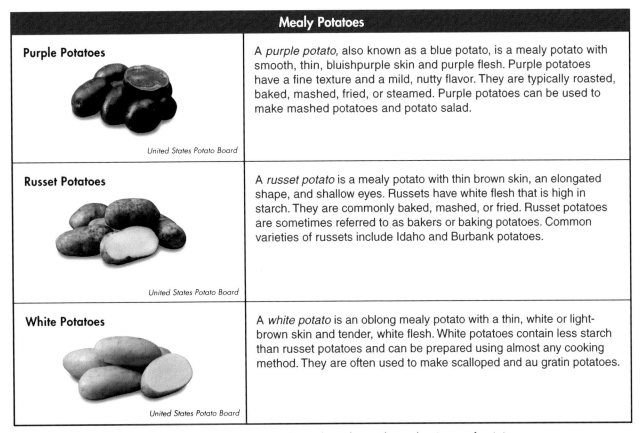

Mealy Potatoes	
Purple Potatoes *United States Potato Board*	A *purple potato,* also known as a blue potato, is a mealy potato with smooth, thin, bluishpurple skin and purple flesh. Purple potatoes have a fine texture and a mild, nutty flavor. They are typically roasted, baked, mashed, fried, or steamed. Purple potatoes can be used to make mashed potatoes and potato salad.
Russet Potatoes *United States Potato Board*	A *russet potato* is a mealy potato with thin brown skin, an elongated shape, and shallow eyes. Russets have white flesh that is high in starch. They are commonly baked, mashed, or fried. Russet potatoes are sometimes referred to as bakers or baking potatoes. Common varieties of russets include Idaho and Burbank potatoes.
White Potatoes *United States Potato Board*	A *white potato* is an oblong mealy potato with a thin, white or light-brown skin and tender, white flesh. White potatoes contain less starch than russet potatoes and can be prepared using almost any cooking method. They are often used to make scalloped and au gratin potatoes.

Figure 3-65. Mealy potatoes are higher in starch and lower in moisture than other types of potatoes.

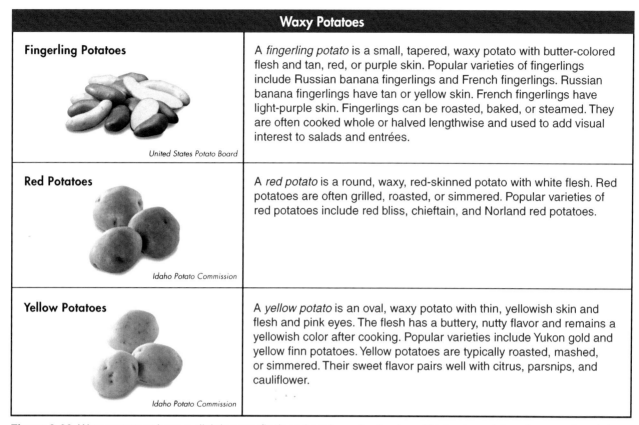

Waxy Potatoes	
Fingerling Potatoes *United States Potato Board*	A *fingerling potato* is a small, tapered, waxy potato with butter-colored flesh and tan, red, or purple skin. Popular varieties of fingerlings include Russian banana fingerlings and French fingerlings. Russian banana fingerlings have tan or yellow skin. French fingerlings have light-purple skin. Fingerlings can be roasted, baked, or steamed. They are often cooked whole or halved lengthwise and used to add visual interest to salads and entrées.
Red Potatoes *Idaho Potato Commission*	A *red potato* is a round, waxy, red-skinned potato with white flesh. Red potatoes are often grilled, roasted, or simmered. Popular varieties of red potatoes include red bliss, chieftain, and Norland red potatoes.
Yellow Potatoes *Idaho Potato Commission*	A *yellow potato* is an oval, waxy potato with thin, yellowish skin and flesh and pink eyes. The flesh has a buttery, nutty flavor and remains a yellowish color after cooking. Popular varieties include Yukon gold and yellow finn potatoes. Yellow potatoes are typically roasted, mashed, or simmered. Their sweet flavor pairs well with citrus, parsnips, and cauliflower.

Figure 3-66. Waxy potatoes have a slightly waxy flesh and are lower in starch and higher in moisture than mealy potatoes.

New Potatoes

United States Potato Board

Figure 3-67. Because they are harvested so early, new potatoes are small and relatively uniform in size.

Sweet Potatoes and Yams

Sweet potatoes and yams have a similar appearance but differ from each other. **See Figure 3-68.** A *sweet potato* is a tuber that grows on a vine and has a paper-thin skin and flesh that ranges in color from ivory to dark orange. Different varieties can be yellow, red, or brown in skin color with yellow to orange-red flesh. The edible skin is often removed before cooking. Sweet potatoes are often incorporated into breads and desserts. Sweet potatoes

can be baked, roasted, sautéed, puréed, or fried. Baking or roasting caramelizes the sugars in sweet potatoes and releases a sweet flavor. Sweet potatoes are an excellent source of vitamin A and potassium.

Potatoes and Yams

Sweet Potato

Yam

Melissa's Produce

Figure 3-68. Sweet potatoes and yams have a similar outer appearance but differ from each other.

A *yam* is a large tuber that has thick, barklike skin and flesh that varies in color from ivory to purple. The flavor of a yam is somewhat dry and starchier than a sweet potato. Common varieties include tropical yams, garnet yams, and jewel yams. Yams are low in fat and a good source of carbohydrates, protein, vitamin A, and vitamin C.

Market Forms of Potatoes

Potatoes can be purchased fresh, frozen, canned, dehydrated, or as processed items such as potato chips. **See Figure 3-69.** Various market forms of potatoes can be used to reduce time and labor costs. For example, frozen precooked French fries, canned cooked whole potatoes, and dehydrated mashed potatoes are often used in the professional kitchen.

Market Forms of Potatoes

United States Department of Agriculture

Figure 3-69. Market forms of potatoes include processed, frozen, dehydrated, and fresh potatoes. Potatoes are also available canned.

Storing Fresh Potatoes

Fresh potatoes are generally sold in 50 lb cases and are packed according to the size of the potato. They may be packaged as 80 count, 90 count, or 100 count. Potatoes must be inspected upon receipt to ensure they are firm and undamaged and show no signs of sprouting or green patches. Potatoes that are purchased cleaned have a shorter

storage life because cleaning removes a protective outer coating that deters bacteria. USDA grading of potatoes is voluntary. U.S. No. 1, U.S. Commercial, and U.S. No. 2 grade potatoes are used in the professional kitchen.

Fresh potatoes must be kept in a dry, cool, dark place that allows them to breathe. Fresh potatoes are best stored in cardboard boxes or mesh bags to allow adequate ventilation. **See Figure 3-70.** The temperature of the storage area should be between 45°F and 55°F. Higher temperatures will shorten the lifespan of the potatoes because potatoes begin to sprout and dehydrate under higher temperatures. Refrigeration is not recommended, as colder temperatures increase the conversion of starch to sugar.

Potato Packaging

Idaho Potato Commission

Figure 3-70. Potatoes are best stored in cardboard boxes or mesh bags to allow adequate ventilation.

Production Tip

Potatoes that are exposed to too much light will turn green and begin to sprout. If potatoes do not have adequate ventilation, they quickly rot.

CHECKPOINT 3-7

1. Describe the four major classifications of potatoes.

2. Identify five market forms of potatoes.

3. Describe the guidelines for receiving and storing potatoes.

COOKING POTATOES

Potatoes can be cooked using a variety of cooking methods and are served for breakfast, lunch, and dinner. Potatoes are often added to soups, braised dishes, and stewed dishes for additional texture, flavor, and nutrients. Potatoes also help thicken sauces by releasing starch during the cooking process. A shorter cooking time requires smaller pieces of potatoes than a dish that cooks for a longer period. If a dish with a longer cooking time calls for small-dice potatoes, the potatoes should be added late in the cooking process to avoid overcooking. Potatoes can be grilled, roasted, baked, sautéed, fried, or simmered.

Production Tip

Uncooked potatoes begin to oxidize immediately when exposed to air, causing them to discolor. To prevent discoloration, potatoes should be placed in cold water as soon as they are peeled or cut.

Grilling Potatoes

Grilled potatoes have a golden color and smoky flavor. **See Figure 3-71.** When grilling potatoes, they are sliced into ¼ inch thick slices or ½ inch wedges, coated with oil and seasonings, and then placed on a preheated grill. Grilled potatoes should be cooked al dente. Overcooking potatoes makes them soft and difficult to remove from the grill without breaking.

Grilled Potatoes

Figure 3-71. Grilled potatoes have a golden color and smoky flavor.

Roasting Potatoes

Potatoes can be roasted or baked. Roasted potatoes are usually cut up, tossed with oil and other seasonings, and then cooked in an oven. Waxy potatoes, such as red, yellow, and fingerling potatoes, can be tossed lightly in oil, seasoned, and then roasted until golden brown on the outside and soft and creamy on the inside. **See Figure 3-72.** Mealy potatoes can also be roasted in this way, but will have a slightly tough exterior. Roasted potatoes are done when a fork can be inserted into the potato with little resistance. Roasted potatoes are typically cooked between 350°F and 400°F for 20 – 45 minutes.

Roasted Potatoes

Figure 3-72. Waxy potatoes can be tossed lightly in oil, seasoned, and then roasted until golden brown.

Cooking Potatoes

Baking Potatoes

Baked potatoes are cooked whole. Russet potatoes make the best baked potatoes. When baked, these low-moisture potatoes produce fluffy and light flesh that readily absorbs butter, sour cream, and a wide variety of other toppings. **See Figure 3-73.** Baked potatoes are typically cooked between 350°F and 400°F until fork tender.

Baked potatoes should be served as quickly as possible because their quality deteriorates if they are held too long. They can be baked directly on the oven rack or can be placed on sheet pans. The skin of a potato is washed thoroughly and pricked with a fork prior to baking. The tiny holes left by the fork allow moisture to escape during the baking process.

Baked Potatoes

United States Potato Board

Plain Baked Potato

Idaho Potato Commission

Toppings

Figure 3-73. When baked, low-moisture potatoes produce fluffy and light flesh that readily absorbs butter, sour cream, and a wide variety of other toppings.

Potato casseroles are also baked. Mealy potatoes are the best choice for potato casseroles because they are low in moisture and can absorb the most flavor from a sauce. A potato casserole, such as scalloped potatoes or potatoes au gratin, can be baked with or without toppings, such as bread crumbs or cheese. **See Figure 3-74.** *Gratinée* is the process of topping a dish with a thick sauce, cheese, or bread crumbs

and then browning it in a broiler or high-temperature oven. *Gratin* is any dish prepared using the gratinée method.

Potatoes used in casseroles may be sliced raw or partially cooked prior to slicing. The sauce used for a casserole can range from a cream sauce to a stock. Using parcooked (partially cooked) potatoes or a warmed sauce will decrease the required baking time.

Potato Casseroles

Scalloped

Au Gratin

Basic American Foods

Figure 3-74. A potato casserole, such as scalloped potatoes or potatoes au gratin, can be baked with or without toppings, such as bread crumbs or cheese.

Procedure for Preparing Potato Casseroles

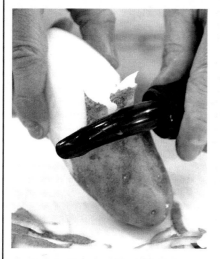

1. Wash and peel the potatoes. Place them in cold water to avoid browning.

2. Slice potatoes to a uniform thickness. If precooked potatoes are desired, steam or simmer them until they are half done. Drain the cooking liquid from the potatoes and allow them to cool slightly.

3. Lay a single layer of the potatoes in a casserole dish.

4. Cover the potatoes with a warm, well-seasoned sauce.

5. Add another layer of potatoes and cover them with sauce. Continue this process until the potatoes and sauce are used.

6. Top with cheese or bread crumbs and cover the pan with foil.

7. Bake at 350°F until potatoes are almost tender. Uncover and allow the potatoes to brown.

Sautéing Potatoes

Waxy potatoes are best for sautéing. Large waxy potatoes are sliced for sautéing and small potatoes are usually cut in half. Sautéed potatoes are typically steamed or boiled first and then finished in a sauté pan to give them a golden-brown exterior. Potatoes that are steamed or boiled first should be removed from the cooking liquid and allowed to steam dry before they are sautéed. Never rinse steamed potatoes to cool them because they absorb even more water and will not brown properly.

Preparing Potato Casseroles
Media Clip

Frying Potatoes

Deep-fried potatoes, such as French fries and potato chips, are a popular side. Fried potatoes can be prepared in a variety of shapes including waffle-cut fries, shoestring fries, steak fries, cottage fries, and hash browns. **See Figure 3-75.** Low-moisture, mealy potatoes, such as russets, are the best choice for deep-frying. They crisp nicely on the outside and remain light and fluffy on the inside. Low-moisture potatoes also spatter less and stay crisp longer.

Fried Potatoes

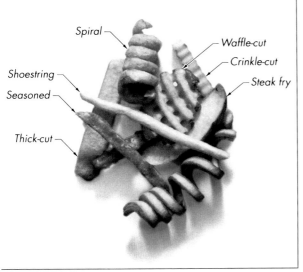

United States Potato Board

Figure 3-75. Fried potatoes can be prepared in a variety of shapes including waffle-cut fries, shoestring fries, steak fries, cottage fries, and hash browns.

Thinly cut potatoes, such as potato chips can be deep-fried in a single step, usually in hot oil between 350°F and 375°F. French fries need to be blanched and cooled before being deep-fried. Blanching potatoes in 275°F oil removes some of the moisture, causing them to cook more evenly when they are deep-fried. Potatoes that are cooked in 375°F oil without being blanched first instantly sear, preventing their interior moisture from escaping. The trapped internal moisture causes the potatoes to become limp and soggy soon after frying.

Simmering Potatoes

Any type of potato can be simmered with good results. If the simmered potatoes are to be mashed, mealy potatoes are the best choice. Waxy potatoes are used when preparing whole, quartered, or sliced potatoes or potato salad because

they retain their shape and do not fall apart as easily. **See Figure 3-76.** Herbs and aromatics can add subtle flavors to simmered potatoes. To add even more flavor when simmering potatoes, stocks or juices can be used as the cooking liquid instead of water.

Simmered Potatoes

United States Potato Board

Mashed

Idaho Potato Commission

Sliced

Figure 3-76. If simmered potatoes are to be mashed, mealy potatoes are best. Waxy potatoes are used for whole, quartered, or sliced potatoes and potato salad because they do not fall apart as easily.

If simmered potatoes are to be served whole, they should be placed on a sheet pan in a single layer to dry. Simmered potatoes have sufficiently dried and are ready to garnish and serve when steam no longer rises from the potatoes. Simmered potatoes are typically seasoned with butter and chopped fresh parsley.

Simmered potatoes can also be puréed before service. To purée simmered potatoes, the drained potatoes are passed through a ricer or food mill until a fine, smooth texture is achieved. Milk, butter, roasted garlic, fresh herbs, grated cheeses, and pesto can be added to simmered potatoes.

Determining Doneness of Potatoes

Fully cooked potatoes are fluffy and have little resistance when pierced by a fork. If a potato is hard in the center, it needs to be cooked longer. Roasted and fried potatoes should be crunchy on the outside and soft on the inside. Cutting open these types of cooked potatoes is the only way to determine if they are done. Lightly squeezing a baked potato will indicate whether it is done. The skin of a fully roasted potato will look slightly shriveled, and the skin of a simmered potato will begin to separate from the flesh when the potato is done.

Plating Potatoes

The potato is not always presented as elegantly as it could be. Too many times cooks only offer baked, mashed, or fried potatoes. The potato of today is a very versatile starch. The addition of bleu cheese and fresh chives to whipped sweet potatoes adds variety, flavor, and contrast to the plate. Russet potatoes and sweet potatoes can be boiled separately and then piped simultaneously to create a swirl of color on the plate. Yukon gold potatoes offer a deep creamy color with great flavor. Mixing new potatoes and Peruvian purple potatoes can add intrigue and loads of flavor. Shape is another important element to keep in mind when choosing potatoes, such as fingerlings, to complement a protein or another vegetable on the plate. Texture variations can be achieved by grilling, baking, roasting, sautéing, frying, or simmering potatoes. **See Figure 3-77.**

CHECKPOINT 3-8

1. Describe six different methods of cooking potatoes.

2. Grill potatoes and evaluate the results.

3. Roast potatoes and evaluate the results.

4. Bake potatoes and evaluate the results.

5. Sauté potatoes and evaluate the results.

6. Fry potatoes and evaluate the results.

7. Simmer potatoes and evaluate the results.

8. Explain how to determine the doneness of potatoes.

Basic American Foods

Plating Potatoes

Idaho Potato Commission

French Fries

United States Potato Board

Steak Fries

Figure 3-77. Shape is an important element to keep in mind when choosing potatoes as a complement.

GRAIN CLASSIFICATIONS

A *grain* is the edible fruit, in the form of a kernel or seed, of a grass. The name of the grass is often the same name given to the grain from that plant. For instance, wheat grass produces the grain called wheat. Grains are often referred to as starches. Common grains include rice, wheat, corn, and barley, but there are many other types of edible grains used in the professional kitchen.

Grain Composition

A kernel of grain is composed of a husk, bran, endosperm, and germ. **See Figure 3-78.** The *husk,* also known as the hull, is the inedible, protective outer covering of grain. The *bran* is the tough outer layer of grain that covers the endosperm. While it is often removed during processing, bran provides necessary fiber, complex carbohydrates, vitamins, and minerals. Studies indicate that bran can help lower cholesterol.

Grain Composition

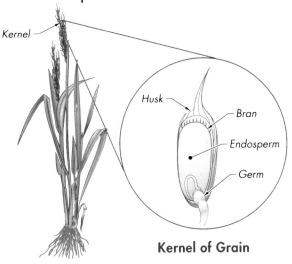

Kernel of Grain

Figure 3-78. A kernel of grain is composed of a husk, bran, endosperm, and germ.

Grains

The *endosperm* is the largest component of a grain kernel and consists of carbohydrates and a small amount of protein. It is milled to produce flours and other grain products. Endosperm consists of carbohydrates (in the form of starches) and a small amount of protein. The *germ* is the smallest part of a grain kernel and contains a small amount of natural oils as well as vitamins and minerals.

Grains come in many shapes and sizes. Most grains are hard and all have a husk that is indigestible in its natural form. It is necessary to process grains to some extent to make them easier to digest. However, the amount of processing directly affects the nutritional values of the grain.

Whole Grains and Refined Grains

A *whole grain* is a grain that only has the husk removed. Because the rest of the grain is intact, whole grains require more time to cook than processed grains. Examples of whole grains include brown rice, wild rice, wheat berries, bulgur, oats, barley, rye, quinoa, millet, and spelt. A *cracked grain* is a whole kernel of grain that has been cracked by being placed between rollers. Bulgur wheat is a cracked grain. Some whole grains are parcooked to reduce the required amount of cooking time.

A *refined grain* is a grain that has been processed to remove the germ, bran, or both. **See Figure 3-79.** Refined grains are easier to digest, have a longer shelf life, can have their color changed, and take less time to cook. Processing often goes beyond removing the husk and can include removing the nutrient-rich bran and germ. The bran is removed from common all-purpose flour to make the product white in color instead of brown. Removing the germ can help preserve the product so that it does not spoil quickly. However, grains lose vitamins, minerals, and fiber during processing.

Refined Grains

United States Department of Agriculture

Figure 3-79. Refined grains have been processed to remove the germ, bran, or both.

Refined grains have been milled, pearled, or flaked. A *milled grain* is a refined grain that has been ground into a fine meal or powder. Meal, such as cornmeal, and all varieties of flour are milled grains. A *pearled grain* is a refined grain with a pearl-like appearance that results from having been scrubbed and tumbled to remove the bran. A *flaked grain,* also known as a rolled grain, is a refined grain that has been rolled to produce a flake. Oatmeal is a flaked grain.

Rice

Rice is a staple food source for two-thirds of the world's population. There are over 40,000 varieties of rice. Rice can be cooked whole or made into flour, noodles, paper, milk, wine, vinegar, and a wide variety of processed foods. Rice is a complex carbohydrate that contains only a trace amount of fat. Rice is cholesterol, sodium, and gluten free and a rich source of thiamine, niacin, iron, phosphorus, and potassium. The high starch content of rice allows it to absorb flavors from ingredients that are cooked with it.

USA Rice Federation

All varieties of rice are milled whole and then rolled to remove the husk. Many varieties are sold as either brown or white rice, depending upon how they were milled. *Brown rice* is rice that has had only the husk removed. Brown rice is chewier and nuttier in flavor than white rice and contains more vitamins and fiber. White rice has had the husk, bran, and germ removed and is more delicate than brown rice. Regardless of its color, rice is classified by the size of the grain. Grain sizes of rice include short-grain, medium-grain, and long-grain. **See Figure 3-80.** The volume of rice triples when it is cooked.

Rice	
Short-Grain Rice  *Indian Harvest Specialtifoods, Inc./Rob Yuretich*	*Short-grain rice* is rice that is almost round in shape and has moist grains that stick together when cooked. It is high in starch and commonly used to make risotto and rice pudding. Arborio rice is a short-grain rice used to make risotto, pilaf, and paella. Sweet rice, also known as sticky rice, is a short-grain rice used to make desserts. It is not used in sushi. Sweet rice is soaked for several hours and then steamed. It is sometimes ground into flour to make dumplings, pastries, or rice puddings.
Medium-Grain Rice  *Indian Harvest Specialtifoods, Inc./Rob Yuretich*	*Medium-grain rice* is a rice that contains slightly less starch than short-grain rice but is still glossy and slightly sticky when cooked. It has a similar appearance to short-grain varieties, with the grains being only slightly longer and plumper. Medium-grain rice is widely used to make sushi. Forbidden rice, also known as emperor's rice, is a black medium-grain rice that turns indigo when cooked. Thai sticky rice is a medium-grain rice that turns a deep-purple color when cooked. Each grain is purple and brown interspersed with flecks of white.
Long-Grain Rice  *Indian Harvest Specialtifoods, Inc./Rob Yuretich*	*Long-grain rice* is a rice that is long and slender and remains light and fluffy after cooking. *Basmati rice* is a long-grain rice that only expands lengthwise when it is cooked. It has a sweet, nutty flavor and is high in fiber. The dry nature of basmati makes it pair well with sauces. Jasmine rice is a long-grain rice that releases a floral aroma when cooked. Himalayan red rice has a rose-colored bran, firm texture, and nutty flavor. Wild rice is not true rice, but its long, black grains are high in nutrients and have a rich, nutty flavor. Wild rice comes from an aquatic grass and is prepared in the same manner as rice.

Figure 3-80. Regardless of its color, rice is classified by the size of the grain.

Corn

Corn, also known as maize, is a cereal grain cultivated from an annual grass that bears kernels on large woody cobs called ears. **See Figure 3-81.** It is prepared as a grain and as a vegetable. Tiny ears of immature corn, also known as baby corn, are often added to stir-fries and salads or eaten whole. Corn is used to make cornstarch, corn oil, corn syrup, and whiskey. Corn is also used to make cornmeal, grits, and hominy.

- Cornmeal is coarsely ground corn. It is commonly used to make cornbread and as a coating for fried foods.

- *Grits* are a coarse type of meal made from ground corn or hominy. Grits are traditionally served with butter, salt, and pepper. Cheese, bacon, or hot sauce may also be added.

- *Hominy* is the hulled kernels of corn that have been stripped of their bran and germ and then dried. White hominy is made from white corn kernels, and yellow hominy is made from yellow corn kernels. Hominy can be ground into flour and used to make tortillas or tamales. Hominy is boiled whole or ground into grits. Hominy is also used to make posole.

Corn

United States Department of Agriculture

Figure 3-81. Corn, also known as maize, is cultivated from an annual grass that bears kernels on large woody cobs called ears.

Wheat

Wheat is a light yellow cereal grain cultivated from an annual grass that yields the flour used in many pastas and baked goods. The most common form of wheat used in cooking is milled flour. Durum wheat, bulgur wheat, and wheat berries are derivatives of wheat that are frequently used in the professional kitchen. **See Figure 3-82.**

Durum Wheat By-Products

Moroccan Couscous

Wheat Berries

Indian Harvest Specialtifoods, Inc./Rob Yuretich

Figure 3-82. Derivatives of wheat are often use in the professional kitchen.

Durum Wheat. *Durum wheat* is the hardest type of wheat. Its high protein content and gluten strength make durum the wheat of choice for making pasta dough. Durum kernels are amber colored and larger than other wheat kernels. Its yellow endosperm gives pasta its golden hue. Semolina and couscous are made from durum wheat.

- *Semolina* is the granular product that results from milling the endosperm of durum wheat. When semolina is mixed with water it forms a stiff dough that can be forced through metal dies to create different pasta shapes.

- *Couscous* is a tiny, round pellet made from durum wheat that has had both the bran and germ removed. It is very fine in texture and similar in size to cornmeal. Israeli couscous is a larger variety.

Bulgur Wheat. Bulgur wheat is golden-brown, nutty-tasting wheat kernels. The husks and bran are removed and it is steamed, dried, and ground. Bulgur comes whole or is cracked into fine, medium, or coarse grains. Bulgur wheat is commonly simmered and seasoned with herbs, spices, and vegetables and cooks in less time than wheat berries. Tabbouleh is made from cooked, chilled bulgur wheat mixed with mint, lemon, olive oil, and parsley.

Wheat Berries. A wheat berry is a chewy wheat kernel with only the husk removed. It contains both the bran and germ. Wheat berries are simmered using a procedure similar to the procedure used for rice. After simmering for an extended period, they can be finished by sautéing them in a pan with various herbs and spices.

Other Grain Varieties

In addition to rice, corn, and wheat there are many other grains used in the professional kitchen. Barley, buckwheat, faro, millet, oats, quinoa, rye, and spelt are only a few of the grains that can be used to add texture, flavor, contrast, and aroma to many dishes. **See Figure 3-83.**

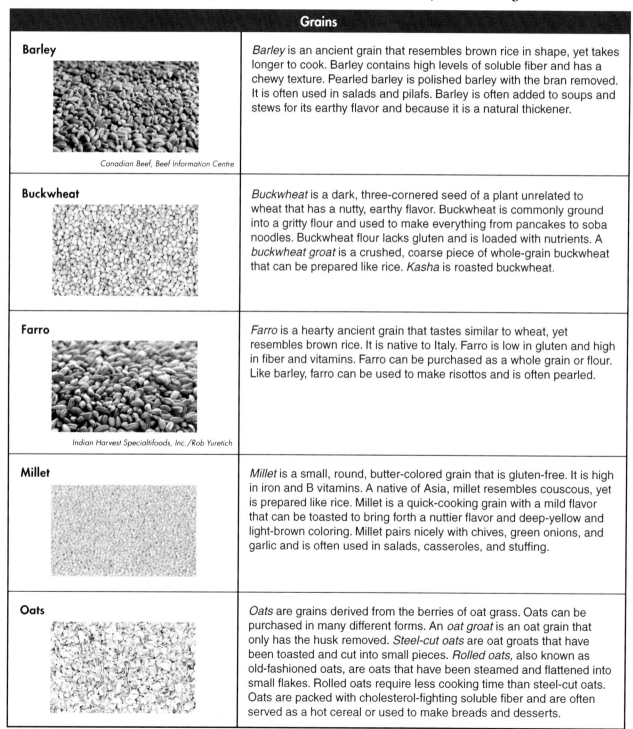

Grains	
Barley *Canadian Beef, Beef Information Centre*	*Barley* is an ancient grain that resembles brown rice in shape, yet takes longer to cook. Barley contains high levels of soluble fiber and has a chewy texture. Pearled barley is polished barley with the bran removed. It is often used in salads and pilafs. Barley is often added to soups and stews for its earthy flavor and because it is a natural thickener.
Buckwheat	*Buckwheat* is a dark, three-cornered seed of a plant unrelated to wheat that has a nutty, earthy flavor. Buckwheat is commonly ground into a gritty flour and used to make everything from pancakes to soba noodles. Buckwheat flour lacks gluten and is loaded with nutrients. A *buckwheat groat* is a crushed, coarse piece of whole-grain buckwheat that can be prepared like rice. *Kasha* is roasted buckwheat.
Farro *Indian Harvest Specialtifoods, Inc./Rob Yuretich*	*Farro* is a hearty ancient grain that tastes similar to wheat, yet resembles brown rice. It is native to Italy. Farro is low in gluten and high in fiber and vitamins. Farro can be purchased as a whole grain or flour. Like barley, farro can be used to make risottos and is often pearled.
Millet	*Millet* is a small, round, butter-colored grain that is gluten-free. It is high in iron and B vitamins. A native of Asia, millet resembles couscous, yet is prepared like rice. Millet is a quick-cooking grain with a mild flavor that can be toasted to bring forth a nuttier flavor and deep-yellow and light-brown coloring. Millet pairs nicely with chives, green onions, and garlic and is often used in salads, casseroles, and stuffing.
Oats	*Oats* are grains derived from the berries of oat grass. Oats can be purchased in many different forms. An *oat groat* is an oat grain that only has the husk removed. *Steel-cut oats* are oat groats that have been toasted and cut into small pieces. *Rolled oats,* also known as old-fashioned oats, are oats that have been steamed and flattened into small flakes. Rolled oats require less cooking time than steel-cut oats. Oats are packed with cholesterol-fighting soluble fiber and are often served as a hot cereal or used to make breads and desserts.

Figure 3-83. (continued on next page)

Grains	
Quinoa *Indian Harvest Specialtifoods, Inc./Rob Yuretich*	*Quinoa* is a small, round, gluten-free grain that is classified as a complete protein. Quinoa is one of the oldest known grains and is available in ivory, red, pink, brown, and black varieties. Quinoa contains fiber, protein, vitamins, and minerals. It is a rich source of the amino acid lysine, which promotes tissue growth and repair and supports the immune system. Quinoa must be rinsed before cooking to remove its bitter coating. It cooks quickly, has a mild flavor, and has a slightly crunchy texture. Quinoa is often used in salads, stuffing, quick breads, and as a side dish.
Rye	*Rye* is a hearty grain with dark-brown kernels that are longer and thinner than wheat. Rye has a distinctive flavor. Rye is used to make flour, breads, crackers, and whiskey. Rye flour is heavier and darker in color than wheat flours. Soaked and cooked rye berries are sometimes added to breads for extra texture or used to make pilafs or hot breakfast cereals. Triticale is a hybrid grain made by crossing rye and wheat. It has a sweet, nutty flavor and contains more protein and less gluten than wheat. Like rye, triticale makes heavy, hearty loaves of bread. Triticale is sold as whole berries, flaked, and flour.
Spelt	*Spelt* is an ancient grain with a nutty flavor and high protein content that is also a good source of riboflavin, zinc, and dietary fiber. Spelt is commonly mistaken for farro because of their similar appearance. Spelt can be used as a hot breakfast cereal or as a substitute for wheat in many dishes. Though it contains gluten, spelt is often tolerated by people with wheat allergies.

Figure 3-83. Barley, buckwheat, farro, millet, oats, quinoa, rye, and spelt can be used to add texture, flavor, contrast, and aroma to many dishes.

Nutrition Note

Grains do not contain any saturated fat or cholesterol. They are a rich source of protein, but the type of protein found in most grains lacks some of the essential amino acids. Carbohydrates account for 65–90% of the calorie content of grains. The bran and germ of whole grains are loaded with vitamins, minerals, and fiber.

Storing Grains

Grains should be stored in a cool, dry place in an airtight container to keep out moisture and prevent insects from getting in the product. **See Figure 3-84.** Some grains can absorb strong aromas, so they should be stored away from foods such as garlic and onions. Grains that still contain germ should be used quickly or kept in a refrigerator or freezer to prevent the germ from becoming rancid. Milled grains will keep indefinitely in sealed containers that are stored in a cool, dry place. Brown rice will last for several months.

Storing Grains

Carlisle FoodService Products

Figure 3-84. Grains should be stored in a cool, dry place in an airtight container to keep out moisture and prevent insects from getting into the product.

CHECKPOINT 3-9

1. Identify the four parts of a whole grain.

2. Differentiate between whole grains and refined grains.

3. Explain how whole grains are cracked.

4. Name three types of refined grains.

5. Explain how grains are milled.

6. Explain how grains are pearled.

7. Explain how grains are flaked.

8. Describe the three major classifications of rice.

9. Identify forms of corn used in the professional kitchen.

10. Identify forms of wheat used in the professional kitchen.

11. Identify forms of oats used in the professional kitchen.

12. Describe barley and how it can be used in the professional kitchen.

13. Describe quinoa and how it can be used in the professional kitchen.

14. Describe how rye can be used in the professional kitchen.

15. Describe buckwheat and how it can be used in the professional kitchen.

16. Describe how farro, millet, and spelt can be used in the professional kitchen.

17. Explain the importance of storing grains in an airtight container and in a cool, dry place.

SIMMERING GRAINS

Grains are most commonly simmered in a hot liquid until all of the liquid has been absorbed by the grain. With the exception of flaked grains and hominy, grains expand in volume when they are cooked. **See Figure 3-85.** The flavors of grains are often enhanced by the liquid that is use to cook them. Stocks, bouillons, consommés, or juices may be used instead of water to add flavor. The addition of aromatic herbs and vegetables heightens the flavor of grains. For example, a pinch of saffron can add flavor and color to grains.

Cooking Grains		
Dry Grain (1 cup)	**Liquid (in cups)**	**Yield* (in cups)**
Arborio rice	2½	2½
Barley, pearled	3	3½–4
Basmati rice, brown	2	3½
Basmati rice, white	1¾	3½
Brown rice, long-grain	2	3½
Brown rice, short-grain	2	3¾
Buckwheat groats, unroasted	2	3½
Bulgur wheat	2	2½–3
Cornmeal polenta	2½	3½
Couscous	1¼	2¼
Forbidden rice	1¾	2¾
Grits	3	3½–4
Hominy	5	3
Jasmine rice	1½	2
Millet	3	5
Oats, steel-cut	4	2
Quinoa	2	4
Rye, berries	3	3
Rye, flakes	3	2½
Spelt, soaked overnight	3½	2½
Sweet rice	2	2
Wheat berries	2½	3
Wheat flakes	4	2
White rice	2	2½

* Yields are approximate

Figure 3-85. With the exception of flaked grains and hominy, grains expand in volume when they are cooked.

Simmering is the most common method of cooking grains. The grain is simply added to a measured amount of boiling cooking liquid. The appropriate amount of cooking liquid depends on the type and amount of grain being cooked. When the water returns to a boil, it is lowered to a simmer and the pot is covered until the grain is fully cooked and the liquid has been absorbed. Risotto and pilaf are two grain preparations that are sautéed and then simmered. **See Figure 3-86.**

Risotto Preparation

Risotto is a classic Italian dish traditionally made by sautéing and then simmering Arborio rice. Risotto is cooked slowly to release the starches from the grain, resulting in a creamy finished product. Risotto preparation begins with lightly sautéing a grain, such as rice. A liquid is then incorporated gradually in small amounts. Garnishes such as vegetables, meat, or shrimp can be added near the end of the cooking process.

Simmered Grains

Tanimura & Antle®

Risotto

National Cancer Institute
Daniel Sone (photographer)

Pilaf

Figure 3-86. Risotto and pilaf are two grain preparations that are sautéed and then simmered.

Procedure for Preparing Risottos

1. Melt fat in a hot saucepan and sweat onions or shallots.

2. Add grain and stir to coat with fat. Sauté until the grain appears translucent.

3. Add white wine and cook until the wine is almost completely reduced.

4. Pour a small amount of stock into the saucepan and continue to stir the grain until all the liquid has been absorbed.

5. Repeat step 4 until the grain is cooked al dente.

Preparing Risottos
Media Clip

Pilaf Preparation

In a pilaf preparation, the flavoring ingredients and grains are sautéed in fat before adding the liquid to prevent clumping. All of the hot liquid or stock is then added along with any seasonings, and the grain is covered and left to simmer until the liquid has been absorbed. A classic pilaf is finished on the stove, although it can also be finished in the oven.

Determining Doneness of Grains

Grains are done when they are tender enough to eat or all of the cooking liquid has been absorbed. Cooked grains that are being held for service should be kept at 140°F or above. Hot grains should be cooled to 70°F within 2 hours and then covered, dated, and refrigerated for no more than seven days. Reheated grains should reach a core temperature of 165°F before being served.

Plating Grains

Grains are simple ingredients that can be used in countless dishes as an ideal accompaniment. There are also numerous grain medley mixtures available that will add more color and texture to the plate. When serving grains like rice or quinoa as an accompaniment for a sauced dish or a stew, the grain should be packed tightly into an oiled mold. The mold is then inverted on the plate and removed, leaving a uniform mound of grain that is visually pleasing. **See Figure 3-87.** Sauces can be added around the mound of grain to broaden the visual appeal.

Plated Grains

House Foods

Figure 3-87. To serve grains such as rice or quinoa as an accompaniment, a mold can be inverted on the plate and removed, leaving a uniform mound of grain that is visually pleasing.

CHECKPOINT 3-10

1. Prepare a rice dish and evaluate the results.

2. Prepare hominy, grits, or polenta and evaluate the results.

3. Prepare a dish using a wheat grain and evaluate the results.

4. Prepare steel-cut oats and rolled oats. Compare the two dishes.

5. Prepare a dish using barley and evaluate the results.

6. Prepare a dish using quinoa and evaluate the results.

7. Prepare a dish using rye or buckwheat and evaluate the results.

8. Prepare a dish using farro, millet, or spelt and evaluate the results.

9. Demonstrate the risotto method of preparing grains.

10. Identify the temperatures at which grains should be held for service, cooled, and reheated.

PASTA CLASSIFICATIONS

Pasta is one of the most versatile food products used in the professional kitchen. *Pasta* is a term for rolled or extruded products made from a dough composed of flour, water, salt, oil, and sometimes eggs. It can be made fresh or purchased in dried or frozen form. Pasta has a mild flavor that does not compete with the flavors of other ingredients or with the sauces that are often added to it. Many pasta dishes, such as macaroni, fettuccine, spaghetti, rigatoni, lasagna, and tortellini, are named for the type of pasta used in the dish. **See Figure 3-88.**

Most pasta is made from wheat flours that contain a very high percentage of gluten, which provides the strength to hold the shape, form, and texture of the dough when cooked. Pasta can be formed into many different shapes and sizes. Ingredients are sometimes added to pasta dough to produce colored pastas. For example, tomato paste produces red pasta, spinach purée produces green pasta, and squid ink produces black pasta.

Pasta-Named Dishes

Macaroni

Fettuccine

Spaghetti

Rigatoni

Lasagna

Tortellini

Barilla America, Inc.

Figure 3-88. Some pasta dishes, such as macaroni, fettuccine, spaghetti, rigatoni, lasagna, and tortellini, are named for the type of pasta used in the dish.

When pasta dough is soft, it can be shaped by rolling it flat and cutting it to the desired size or by forcing it through the metal dies of a pasta machine to create various shapes. **See Figure 3-89.** The dough is then allowed to dry, resulting in a hard, mold-resistant product. The shape of the pasta chosen for a given dish is often determined by the sauce and how it will cling to the particular pasta shape. In addition, the pasta shape should complement the appearance of the finished dish. Pastas are classified as shaped, tube, ribbon, and stuffed pastas.

Shaped Pastas

A *shaped pasta* is a pasta that has been extruded into a complex shape such as a corkscrew, bowtie, shell, flower, or star. **See Figure 3-90.** Shaped pastas are used in soups, salads, casseroles, and stir-fries. They add visual interest and texture. Common types of shaped pastas include campanelle, conchiglie, farfalle, fiori, fusilli, gemelli, jumbo shells, orecchiette, orzo, radiatori, rotini, and stelline.

Pasta Machine

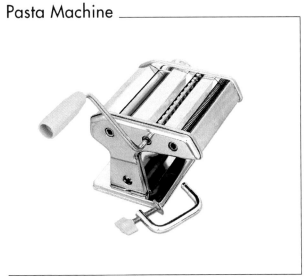

Browne-Halco (NJ)

Figure 3-89. Dough can be fed through a pasta machine to create various shapes of pasta.

Shaped Pastas	
Campanelle *Barilla America, Inc.*	Campanelle, also known as gigli, are shaped pasta that have been rolled into a fluted-cone shape.
Conchiglie (Large Shells) *Barilla America, Inc.*	Conchiglie, also known as large shells, are shell-shaped pasta.
Farfalle *Barilla America, Inc.*	Farfalle, also known as bow tie pasta, are flat squares of shaped pasta that are pinched in the center in the shape of bow ties.
Fiori *Barilla America, Inc.*	Fiori are flower-shaped pasta.
Fusilli *Barilla America, Inc.*	Fusilli are corkscrew-shaped pasta.
Gemelli *Barilla America, Inc.*	Gemelli is formed by twisting two strands of pasta together.
Jumbo Shells *Barilla America, Inc.*	Jumbo shells are large, shell-shaped pasta.
Orecchiette *Barilla America, Inc.*	Orecchiette are small, ridged, bowl-shaped pasta.
Orzo *Barilla America, Inc.*	Orzo are small oval-shaped pasta with an appearance similar to that of a grain.

Figure 3-90. (continued on next page)

Shaped Pastas	
Radiatori  *Barilla America, Inc.*	Radiatori are deeply-ridged, curly-shaped pasta.
Rotini *Barilla America, Inc.*	Rotini are 2 inch long, twist-shaped pasta.
Stelline  *Barilla America, Inc.*	Stelline are tiny, star-shaped pasta.

Figure 3-90. Shaped pastas are made into complex shapes, such as corkscrews, shells, and stars, using an extruder.

Tube Pastas

A *tube pasta* is a pasta that has been pushed through an extruder and then fed through a cutter that cuts the tubes to desired length. **See Figure 3-91.** Tube pastas are often stuffed with cheese or meat and are often used in casseroles. Short tubes are often used in soups. Cellentani, ditalini, elbows, manicotti, penne, pipettes, rigatoni, and ziti are common types of tube pasta.

Photo Courtesy of Perdue Foodservice, Perdue Farms Incorporated

Tube Pastas	
Cellantani  *Barilla America, Inc.*	Cellentani, also known as cavatappi, are twisted, hollow tubes of pasta with a ribbed surface.
Ditalini 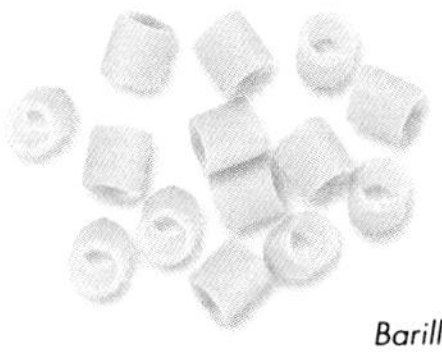 *Barilla America, Inc.*	Ditalini are short, hollow tubes of pasta with a smooth or ridged surface.
Elbows  *Barilla America, Inc.*	Elbows are relatively short, slightly curved, hollow tubes of pasta.
Manicotti *Barilla America, Inc.*	Manicotti, also known as cannelloni, are large, round tubes of pasta approximately 4 inches long and 1 inch in diameter. They can be straight cut or diagonal cut.

Figure 3-91. (continued on next page)

Tube Pastas	
Penne *Barilla America, Inc.*	Penne are hollow, diagonally cut tubes of pasta approximately 1½–2 inches in length with a smooth or ribbed surface.
Pipettes  *Barilla America, Inc.*	Pipettes are curved, hollow, ridged tubes of pasta with one open end and one pinched end.
Rigatoni *Barilla America, Inc.*	Rigatoni are wide, hollow, ridged tubes of pasta.
Ziti  *Barilla America, Inc.*	Ziti are straight, round tubes of pasta approximately ¼ inch in diameter of various lengths with a smooth or ribbed surface.

Figure 3-91. Tube pastas are pushed through a tube-shaped pasta extruder and then fed through a tube cutter that cuts the tubes to desired length.

Ribbon Pastas

A *ribbon pasta* is a thin, round strand or flat, ribbonlike strand of pasta. **See Figure 3-92.** To form ribbon pasta, the dough is rolled out and cut to the desired width by hand or using a pasta machine. Round strands of pasta are formed by passing the dough through small, round dies. Ribbon pastas are often dressed in sauces. Common types of ribbon pasta include capellini, egg noodles, fettuccine, lasagna, linguine, and spaghetti.

Ribbon Pastas	
Capellini *Barilla America, Inc.*	Capellini, also known as angel hair, are very fine, round, strandlike ribbons of pasta approximately 1/64 inch in diameter.
Egg Noodles  *Barilla America, Inc.*	Egg noodles are flat, ribbons of pasta that can be cut long or short and thin, medium, or wide. To be labeled egg noodles, the pasta must contain at least 5.5% egg solids.
Fettuccine *Barilla America, Inc.*	Fettuccine are long, thin, flat strips of ribbon pasta approximately ¼ inch wide.

Figure 3-92. (continued on next page)

Ribbon Pastas	
Lasagna 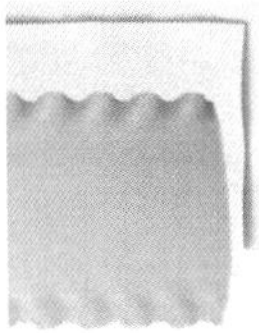 *Barilla America, Inc.*	Lasagna are flat, ripple-edged ribbons of pasta, approximately 2 inches wide.
Linguine  *Barilla America, Inc.*	Linguine are long, thin, flat strips of ribbon pasta, about ⅛ inch wide.
Spaghetti *Barilla America, Inc.*	Spaghetti are long, round rods of ribbon pasta approximately $\frac{3}{32}$ inch in diameter. Very thin strands of spaghetti are known as spaghettini.

Figure 3-92. Ribbon pasta is a thin, round strand or a flat, ribbonlike strand of pasta.

Stuffed Pastas

A *stuffed pasta* is a pasta that has been formed by hand or machine to hold fillings. **See Figure 3-93.** Individual portions of filling are added to a sheet of pasta and then a wash is applied around each filling. Another sheet of pasta is laid over the first sheet and the mounds of filling, the two sheets are sealed with the mounds of filling inside, and then the pasta is cut into individual portions. Stuffed pastas can be filled with savory and sweet cheeses, puréed meats, poultry, seafood, or vegetables or a combination of ingredients made into a paste. Common types of stuffed pasta include ravioli, tortellini, and tortelloni.

Stuffed Pastas	
Ravioli  *Barilla America, Inc.*	Ravioli are stuffed pasta formed from equal size squares, or other shapes, of flat pasta and are filled.
Tortellini 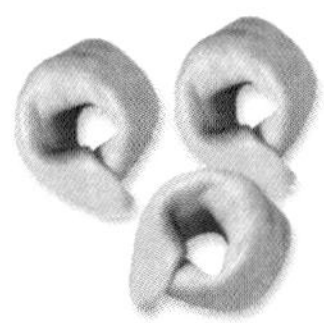 *Barilla America, Inc.*	Tortellini are stuffed pasta formed by wrapping filled half circles of dough around a finger and pressing the ends together.
Tortelloni  *Barilla America, Inc.*	Tortelloni are stuffed pasta that resemble pot stickers, are larger than tortellini, and will hold more filling than tortellini.

Figure 3-93. Ravioli, tortellini, and tortelloni are stuffed pastas.

Asian Noodles

Some Asian noodles are wheat-based, yet many are made from vegetable starch, eggs, rice flour, or buckwheat flour. **See Figure 3-94.** Asian noodles are soaked briefly in hot water rather than cooked. Asian noodles may also be steamed, stir-fried, and deep-fried. Some Asian noodles are served cold as well as hot.

Asian Noodles

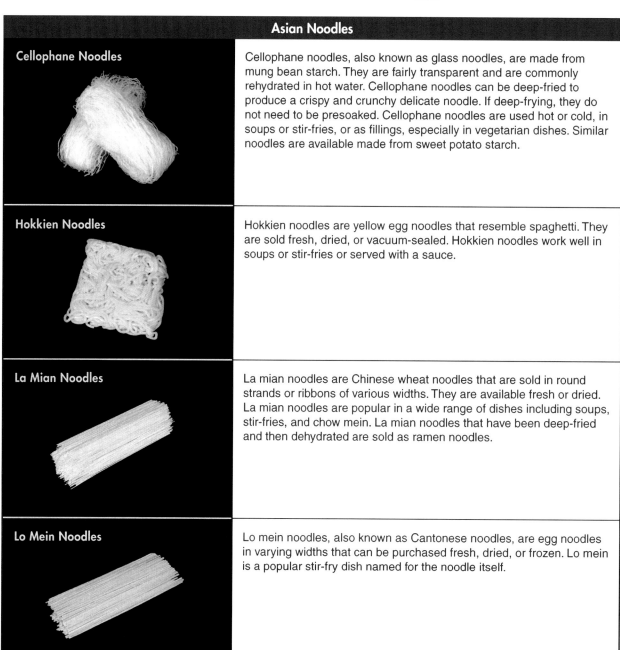

Asian Noodles	
Cellophane Noodles	Cellophane noodles, also known as glass noodles, are made from mung bean starch. They are fairly transparent and are commonly rehydrated in hot water. Cellophane noodles can be deep-fried to produce a crispy and crunchy delicate noodle. If deep-frying, they do not need to be presoaked. Cellophane noodles are used hot or cold, in soups or stir-fries, or as fillings, especially in vegetarian dishes. Similar noodles are available made from sweet potato starch.
Hokkien Noodles	Hokkien noodles are yellow egg noodles that resemble spaghetti. They are sold fresh, dried, or vacuum-sealed. Hokkien noodles work well in soups or stir-fries or served with a sauce.
La Mian Noodles	La mian noodles are Chinese wheat noodles that are sold in round strands or ribbons of various widths. They are available fresh or dried. La mian noodles are popular in a wide range of dishes including soups, stir-fries, and chow mein. La mian noodles that have been deep-fried and then dehydrated are sold as ramen noodles.
Lo Mein Noodles	Lo mein noodles, also known as Cantonese noodles, are egg noodles in varying widths that can be purchased fresh, dried, or frozen. Lo mein is a popular stir-fry dish named for the noodle itself.

Figure 3-94. (continued on next page)

Vegetable, Starch & Pasta Station 83

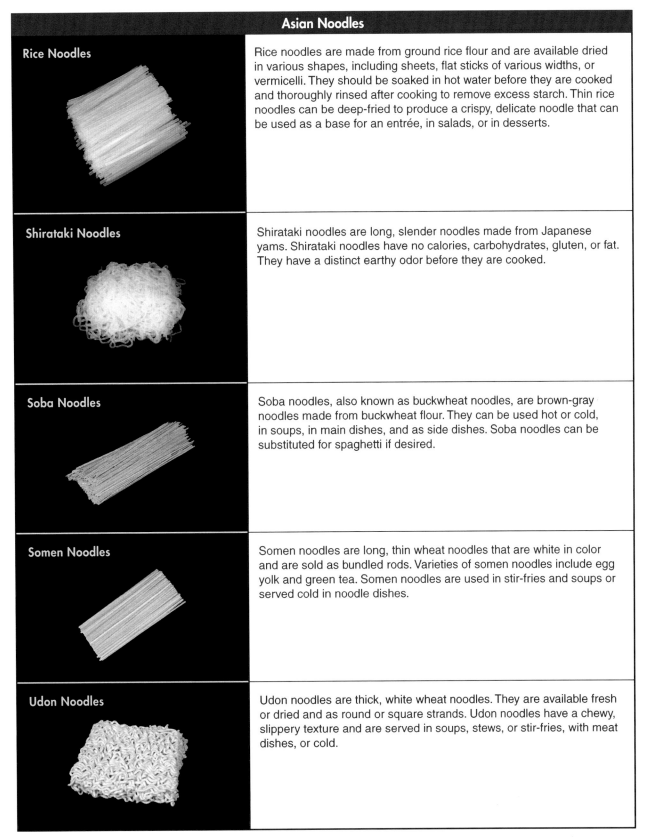

Asian Noodles	
Rice Noodles	Rice noodles are made from ground rice flour and are available dried in various shapes, including sheets, flat sticks of various widths, or vermicelli. They should be soaked in hot water before they are cooked and thoroughly rinsed after cooking to remove excess starch. Thin rice noodles can be deep-fried to produce a crispy, delicate noodle that can be used as a base for an entrée, in salads, or in desserts.
Shirataki Noodles	Shirataki noodles are long, slender noodles made from Japanese yams. Shirataki noodles have no calories, carbohydrates, gluten, or fat. They have a distinct earthy odor before they are cooked.
Soba Noodles	Soba noodles, also known as buckwheat noodles, are brown-gray noodles made from buckwheat flour. They can be used hot or cold, in soups, in main dishes, and as side dishes. Soba noodles can be substituted for spaghetti if desired.
Somen Noodles	Somen noodles are long, thin wheat noodles that are white in color and are sold as bundled rods. Varieties of somen noodles include egg yolk and green tea. Somen noodles are used in stir-fries and soups or served cold in noodle dishes.
Udon Noodles	Udon noodles are thick, white wheat noodles. They are available fresh or dried and as round or square strands. Udon noodles have a chewy, slippery texture and are served in soups, stews, or stir-fries, with meat dishes, or cold.

Figure 3-94. Some Asian noodles are wheat-based, yet many are made from eggs, rice flour, or buckwheat flour.

Storing Pasta

Pasta is often purchased ready to use. Purchased pasta should be placed in an airtight container and stored in a cool, dry place to keep moisture out and prevent insects from getting in the product.

CHECKPOINT 3-11

1. Identify three ways pasta can be purchased.

2. Describe shaped pastas.

3. Describe tube pastas.

4. Describe ribbon pastas.

5. Describe stuffed pastas.

6. Describe nine types of Asian noodles.

PREPARING AND COOKING PASTAS

Most pastas are boiled or baked. Pasta at least doubles in volume when cooked. For example, a pound of fresh pasta yields approximately 2–2½ lb of cooked pasta. Fresh pasta requires less cooking time than dried pasta. Cooking time varies depending on the shape, size, and quality of the pasta. **See Figure 3-95.** Pasta should always be cooked uncovered. After the minimum cooking time is reached, the pasta should be tested frequently to ensure the proper doneness. Fresh pasta cooks in approximately 3–5 minutes, while dried pasta cooks in approximately 3–13 minutes.

Approximate Cooking Times for Dried Pasta			
Pasta	*Minutes*	*Pasta*	*Minutes*
Capellini	3–5	Spaghetti	9–10
Egg Noodles	7–8	Penne and Mostaccioli	9–11
Elbow Macaroni	7–8		
Conchiglie	8–9	Tortellini	10–11
Lasagna	8–9	Farfalle	11–12
Manicotti and Connelloni	8–9	Fettucini	12–13
		Fusilli	12–13
Linguine	9–10	Jumbo shells	12–13
Orzo	9–10	Vermicelli	12–13

Figure 3-95. Cooking time varies depending on the shape, size, and quality of the pasta.

Preparing Pasta Doughs

Some chefs prefer to make their own pasta because fresh pasta is more tender and can be flavored and colored as desired. For example, squid ink can provide a salty, metallic flavor that will also color the pasta. Herbs, spices, and vegetables may also be added to enhance the flavor and appearance of pasta. Fresh pasta dough can be processed in a mixer or kneaded by hand.

Procedure for Preparing Pasta Dough

1. On a clean work surface, place flour in a mound and form a well in the center. Add eggs, oil, and salt to the well.
2. Slowly work the flour into the well with the egg mixture until all ingredients are mixed.

3. Knead the mixture until a smooth, dry ball of dough has formed. Cover the dough and allow it to rest for 20 minutes.

4. Roll the dough to the appropriate thickness and cut into desired shapes.

Preparing Pasta Dough
Media Clip

Preparing Ravioli. To make ravioli, one sheet of pasta is topped with a small amount of filling, and a second sheet of pasta is laid on top. The top sheet is pressed down around the filling, and individual ravioli are formed by cutting around the mounds of filling with a pastry wheel, ravioli cutter, or knife.

Procedure for Preparing Ravioli

1. Roll out a 12 inch, square sheet of pasta dough as thin as possible.

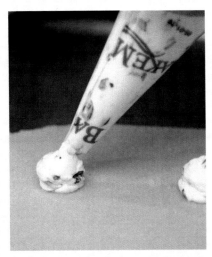

2. Use a pastry bag to deposit ¼ oz portions of filling on the dough, approximately 2 inches apart.

3. Use a pastry brush to egg wash the pasta around each portion of filling.

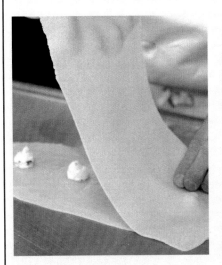

4. Prepare a second sheet of dough the same size as the first and place it on top of the first sheet.

5. Press down around each portion of filling to seal all edges of the two sheets of dough together.

6. Use a pastry wheel or knife to cut between the filled mounds.

7. Separate the ravioli and place on a sheet pan. Cook or cover and refrigerate or freeze for future use.

Preparing Ravioli
Media Clip

Preparing Tortellini. When making tortellini, thin pasta dough is cut into circles and a filling is placed in the center of each circle. The circle is then folded in half and the edges are pressed together to hold the filling in place. The half circle is then formed into a ring by wrapping it around a finger and pressing the two ends together.

Procedure for Preparing Tortellini

1. Roll out pasta dough as thin as possible.

2. Use a 2 inch cutter to cut the dough into rounds.

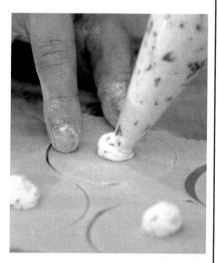

3. Place a small portion of filling in the center of each round.

4. Moisten the edge of each round with water or egg wash.

5. Fold each circle in half and then press the edges together until they are tightly sealed.

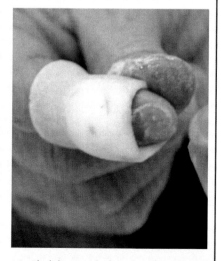

6. Slightly stretch the tips of each half circle to form a ring of dough around a finger. Press the tips together until they are tightly sealed. Cook or cover and refrigerate or freeze.

Determining Doneness of Pastas

Pasta should be cooked al dente, meaning "to the tooth." When pasta is cooked al dente, there should be a slight resistance in the center of the pasta when it is chewed. Pasta should not be undercooked and brittle nor overdone and mushy. For example, rigatoni tubes should not collapse and rotini should not unwind.

Plating Pastas

The ideal pasta plate has a wide, flat bottom and shallow sides that keeps the pasta warm and unruly noodles in their place while supporting the stacking of pasta for visual appeal. Pasta should only be tossed with enough sauce to coat, moisten, and flavor each piece. The pasta should not drown in the sauce. A pasta fork or spaghetti tongs are used to grab the pasta without crushing it and transfer it to the warmed plate. One to two generous tablespoons of sauce are then spooned over the pasta before it is topped with freshly grated cheese or a splash of extra virgin olive oil. **See Figure 3-96.** Pasta is often garnished with chopped parsley.

Plated Pasta

Barilla America, Inc.

Figure 3-97. Pasta is often topped with grated cheese.

Reheating Pasta

In the professional kitchen, pasta is often cooked in advance. When pasta is cooked in advance, it is rinsed in cold water to stop the cooking process, then it is tossed with a small amount of oil to prevent sticking, and stored. Parcooked pasta can be rinsed in warm water or dropped in simmering water to bring it to the temperature needed for service.

CHECKPOINT 3-12

1. Explain how pasta reacts to the cooking process.

2. Prepare fresh pasta dough and evaluate the results.

3. Prepare a tubed pasta dish and evaluate the results.

4. Prepare a ribbon pasta dish and evaluate the results.

5. Prepare a shaped pasta dish and evaluate the results.

6. Prepare a formed pasta dish and evaluate the results.

7. Prepare a dish using vegetable noodles and evaluate the results.

8. Prepare a dish using egg noodles and evaluate the results.

9. Prepare a dish using rice noodles and evaluate the results.

10. Prepare a dish using buckwheat noodles and evaluate the results.

11. Prepare a dish using wheat noodles and evaluate the results.

12. Explain how to prepare ravioli.

13. Explain how to prepare tortellini.

14. Explain how to determine if pasta is cooked al dente.

15. Reheat parcooked pasta and evaluate the results.

Barilla America, Inc.

KEY TERMS

Refer to CD-ROM for **Flash Cards**

- **acorn squash:** A winter squash that looks like a large, dark-green acorn.
- **aggregate fruit:** A cluster of very tiny fruits.
- **Anjou pear:** A plump, lopsided pear that is green in color.
- **apple:** A hard, round pome that can range in flavor from sweet to tart and in color from pale yellow to dark red.
- **apricot:** A drupe that has pale orange-yellow skin with a fine, downy texture and a sweet and aromatic flesh.
- **artichoke:** The edible flower bud of a large, thistle-family plant that comes in many varieties.
- **Asian pear:** A round pear with the texture of an apple and yellow-colored skin. Also known as an apple-pear.
- **asparagus:** A green, white, or purple edible stem vegetable that is referred to as a spear.
- **avocado:** A pear-shaped drupe with a rough green skin and a large pit surrounded by yellow-green flesh. Also known as an alligator pear.
- **banana:** A yellow, elongated tropical fruit that grows in hanging bunches on a banana plant.
- **barley:** An ancient grain that resembles brown rice in shape, yet takes longer to cook.
- **Bartlett pear:** A large, golden pear with a bell shape.
- **basmati rice:** A long-grain rice that only expands lengthwise when it is cooked.
- **batonnet cut:** A stick cut that produces a stick-shaped item ¼ × ¼ × 2 inches long.
- **bean:** The edible seed of various plants in the legume family.
- **beet:** A round root vegetable that is a deep reddish purple or gold color.
- **beet green:** The green, edible leaf that grows out of the top of the beet root vegetable.
- **Belgian endive:** A leaf vegetable with a slender, tightly packed, elongated head that forms a point.
- **bell pepper:** A fruit-vegetable with three or more lobes of crisp flesh that surround hundreds of seeds in an inner cavity.
- **berry:** A type of fruit that is small and has many tiny, edible seeds.
- **blackberry:** A sweet, dark-purple to black, aggregate fruit that grows on a bramble bush.
- **blanching:** A moist-heat cooking technique in which food is briefly parcooked and then shocked by placing it in ice-cold water to stop the cooking process.
- **blueberry:** A small, dark-blue berry that grows on a shrub.
- **bok choy:** A leaf vegetable that has tender white ribs, bright-green leaves, and a more subtle flavor than head cabbage.
- **bolster:** A thick band of metal located where the blade of a knife joins the handle.
- **boning knife:** A thin knife with a pointed 6–8 inch blade used to separate meat from bones with minimal waste.
- **Bosc pear:** A pear with a gourd-like shape and a brown-colored to bronze-colored peel.
- **boysenberry:** A deep-maroon, hybrid berry made by crossbreeding a raspberry, a blackberry, and a loganberry.
- **bran:** The tough outer layer of grain that covers the endosperm.
- **breadfruit:** An exotic fruit that is native to Polynesia and has bumpy green skin and a white starchy flesh.
- **bread knife:** A knife with a serrated blade 8–12 inches long that is used to cut through the crusts of breads without crushing the soft interior.
- **broccoli:** An edible flower that is a member of the cabbage family and has tight clusters of dark-green florets on top of a pale-green stalk with dark-green leaves.
- **brown rice:** Rice that has had only the husk removed.
- **brunoise cut:** A dice cut that produces a cube-shaped item with six equal sides measuring ⅛ inch each.
- **Brussels sprout:** A very small round head of tightly packed leaves that looks like a tiny cabbage.
- **buckwheat:** A dark, three-cornered seed of a plant unrelated to wheat that has a nutty, earthy flavor.
- **buckwheat groat:** A crushed, coarse piece of whole-grain buckwheat that can be prepared like rice.
- **bulb vegetable:** A strongly flavored vegetable that grows underground and consists of a short stem base with one or more buds that are enclosed in overlapping membranes or leaves.
- **butcher's knife:** A heavy knife with a curved tip and a blade that is 7–14 inches in length.
- **butterhead lettuce:** A lettuce with pale-green leaves and a sweet, buttery flavor, tender texture, and delicate structure.
- **butternut squash:** A large, bottom-heavy, tan-colored winter squash.
- **button mushroom:** A cultivated mushroom with a very smooth, rounded cap and completely closed gills atop a short stem. Also known as a white mushroom.
- **canary melon:** A fairly large, bright yellow melon with a smooth rind that is slightly waxy when ripe.
- **cantaloupe:** An orange-fleshed melon with a rough, deeply grooved rind.

KEY TERMS (continued)

- **capsaicin:** A potent compound that gives chiles their hot flavor.
- **carotenoid:** An organic pigment found in orange or yellow vegetables.
- **carrot:** An elongated root vegetable that is rich in vitamin A.
- **casaba melon:** A teardrop-shaped melon with a thick, bright-yellow, ridged rind and white flesh.
- **cauliflower:** An edible flower that is a member of the cabbage family and has tightly packed white florets on a short, white-green stalk with large, pale-green leaves.
- **celeriac:** A knobby, brown root vegetable cultivated from a type of celery grown for its root rather than its stalk. Also known as celery root.
- **celery:** A green stem vegetable that has multiple stalks measuring 12–20 inches in length.
- **channel knife:** A special cutting tool with a thin metal blade within a raised channel that is used to remove a large string from the surface of a food item.
- **chanterelle mushroom:** A trumpet-shaped mushroom that ranges in color from bright yellow to orange and has a nutty flavor and a chewy texture.
- **chard:** A large, dark-green, leaf vegetable with white or reddish stalks.
- **chef's knife:** A large and very versatile knife with a tapering blade used for slicing, dicing, and mincing. Also known as a French knife.
- **cherry:** A small, smooth-skinned drupe that grows in a cluster on a cherry tree.
- **chicory:** A curly leaf vegetable with a slightly bitter-tasting flavor. Also known as escarole.
- **chiffonade cut:** A slicing cut that produces thin shreds of leafy greens or herbs.
- **chile:** A brightly colored fruit-vegetable pod with distinct mild to hot flavors.
- **chopping:** Rough-cutting an item so that there are relatively small pieces throughout, although there is no uniformity in shape or size.
- **clam knife:** A small knife with a short, flat, round-tipped sharp blade that is used to open clams.
- **cleaver:** A heavy, rectangular-bladed knife that is used to cut through bones and thick meat.
- **chlorophyll:** An organic pigment found in green vegetables.
- **citrus fruit:** A type of fruit with a brightly colored, thick rind and pulpy, segmented flesh that grows on trees in warm climates.
- **clementine:** A citrus fruit that is similar to a tangerine, but with a rougher skin.

- **coconut:** A tropical drupe with a white meat that is housed within a hard, fibrous brown husk.
- **collard:** A large, dark-green, edible leaf with a thick, white vein. Also known as a collard green.
- **Comice pear:** A fairly large pear with a rotund body and a very short, well-defined neck.
- **Concord grape:** A seeded grape with a deep black color.
- **couscous:** A tiny, round pellet made from durum wheat that has had both the bran and germ removed.
- **cracked grain:** A whole kernel of grain that has been cracked by being placed between rollers.
- **cranberry:** A small, red, round berry that has a tart flavor.
- **Crenshaw melon:** A large, pear-shaped melon with a yellow-green, slightly ribbed rind and an orange or salmon-colored flesh.
- **crisphead lettuce:** A lettuce with a large, round, tightly packed head of pale-green leaves.
- **cucumber:** A green, cylindrical fruit-vegetable that has an edible skin, edible seeds, and a moist flesh.
- **currant:** A small red, black, or golden-white berry that grows in grape-like clusters.
- **dandelion green:** The dark-green edible leaf of the dandelion plant.
- **date:** A plump, juicy, and meaty drupe that grows on a date palm tree.
- **diagonal cut:** A slicing cut that produces flat-sided, oval slices.
- **dice cuts:** Precise cubes cut from uniform stick cuts.
- **dragon fruit:** An exotic fruit of a cactus with an inedible pink skin, green scales, and white or red flesh speckled with small crunchy black seeds.
- **dried fruit:** Fruit that has had most of the moisture removed either naturally or through the use of a machine, such as a food dehydrator.
- **drupe:** A type of fruit that contains one hard seed or pit. Also known as a stone fruit.
- **dulse:** A stringy, reddish-brown sea vegetable with a fishy odor that is rich in iron, iodine, potassium, and vitamin A.
- **durian:** An exotic fruit that contains several pods of sweet, yellow flesh and has a custard-like texture.
- **durum wheat:** The hardest type of wheat.
- **edamame:** Green soybeans housed within a fibrous, inedible pod.
- **edge:** The sharpened part of the knife blade that extends from the heel to the tip.
- **edible flowers:** The flowers of nonwoody plants that are prepared as vegetables.
- **eggplant:** A deep-purple, white, or variegated fruit-vegetable with edible skin and a yellow to white, spongy flesh that contains small, brown, edible seeds.

KEY TERMS (continued)

- **endosperm:** The largest component of a grain kernel and consists of carbohydrates and a small amount of protein.
- **enokitake mushroom:** A crisp, delicate mushroom that has spaghetti-like stems topped with white caps. Also known as an enoki or a snow puff mushroom.
- **exotic fruit:** Type of fruit that comes from a hot, humid location but is not as readily available as a tropical fruit.
- **farro:** A hearty grain that tastes similar to wheat, yet resembles brown rice.
- **fennel:** A celery-like stem vegetable with overlapping leaves that grow out of a large bulb at its base.
- **fiddlehead fern:** The curled tip of an ostrich fern frond that has a nutty and slightly bitter flavor similar to asparagus and artichokes.
- **fig:** The small pear-shaped fruit of the fig tree.
- **fine brunoise cut:** A dice cut that produces a cube-shaped item with six equal sides measuring 1/16 inch.
- **fine julienne cut:** A stick cut that produces a stick-shaped item 1/16 × 1/16 × 2 inches long.
- **fingerling potato:** A small, tapered, waxy potato with butter-colored flesh and tan, red, or purple skin.
- **flaked grain:** A refined grain that has been rolled to produce a flake. Also known as a rolled grain.
- **flavonoid:** An organic pigment found in purple, dark-red, and white vegetables.
- **fluted cut:** A specialty cut that leaves a spiral pattern on the surface of an item by removing only a sliver with each cut.
- **freeze drying:** The process of removing the water content from a food and replacing it with a gas.
- **frisée:** A leaf vegetable with twisted, thin leaves that grow in a loose bunch. Also known as curly endive.
- **fruit:** The edible, ripened ovary of a flowering plant that usually contains one or more seeds.
- **fruit-vegetable:** A botanical fruit that is sold, prepared, and served as a vegetable.
- **garlic:** A bulb vegetable made up of several small cloves that are enclosed in a thin, husklike skin.
- **germ:** The smallest part of a grain kernel; contains a small amount of natural oils as well as vitamins and minerals.
- **gooseberry:** A smooth-skinned berry that can be green, golden, red, purple, or white and has many tiny seeds on the inside.
- **grain:** The edible fruit, in the form of a kernel or seed, of a grass.
- **grape:** An oval fruit that has a smooth skin and grows on woody vines in large clusters.
- **grapefruit:** A round citrus fruit with a thick, yellow outer rind and tart flesh.

- **gratin:** Any dish prepared using the gratinée method.
- **gratinée:** The process of topping a dish with a thick sauce, cheese, or bread crumbs and then browning it in a broiler or high-temperature oven.
- **grits:** A coarse type of meal made from ground corn or hominy.
- **guava:** A small oval-shaped tropical fruit, usually 2–3 inches in diameter, with thin edible skin that can be yellow, red, or green.
- **head cabbage:** A tightly packed, round head of overlapping leaves that can be green, purple, red, or white in color.
- **heart of palm:** A slender, white, stem vegetable that is surrounded by a tough husk.
- **heel:** The rear portion of a knife blade; most often used to cut thick items where more force is required.
- **hominy:** The hulled kernels of corn that have been stripped of their bran and germ and then dried.
- **honeydew melon:** A melon with a smooth outer rind that changes from a pale-green color to a creamy-yellow color as it ripens.
- **honing:** The process of aligning a blade's edge and removing any burrs or rough spots on the blade. Also known as truing.
- **husk:** The inedible, protective outer covering of grain. Also known as the hull.
- **hybrid fruit:** A fruit that is the result of crossbreeding two or more fruits of different species to obtain a completely new fruit.
- **irradiation:** The process of exposing food to low doses of gamma rays in order to destroy deadly organisms such as E. coli O157:H7, campylobacter, and salmonella.
- **jackfruit:** An enormous, spiny, oval exotic fruit with yellow flesh that tastes like a banana and has seeds that can be boiled or roasted and then eaten.
- **jicama:** A large, brown root vegetable that ranges in size from 4 oz to 6 lb.
- **julienne cut:** A stick cut that produces a stick-shaped item 1/8 × 1/8 × 2 inches long.
- **kale:** A large, frilly, leaf vegetable that varies in color from green and white to shades of purple.
- **kasha:** Roasted buckwheat.
- **key lime:** A variety of lime that is smaller, more acidic, and more strongly flavored than other limes.
- **kiwano:** An exotic fruit with jagged peaks rising from an orange and red-ringed rind that is native to Africa. Also known as horned melon.
- **kiwifruit:** A small, barrel-shaped tropical fruit with a thin, brown, fuzzy skin.

KEY TERMS (continued)

- **kohlrabi:** A sweet, crisp, stem vegetable that has a pale-green or purple, bulbous stem and dark-green leaves.
- **kombu:** A long, dark-brown to purple sea vegetable that is used to flavor dashi stock.
- **kumquat:** A small, golden, oval-shaped exotic fruit with a thin, sweet peel and tart center.
- **leaf vegetables:** Plant leaves that are often accompanied by edible leafstalks and shoots. Also known as greens.
- **leek:** A long, white bulb vegetable, with long, wide, flat leaves.
- **legume:** Edible seed of a nonwoody plant that grows in multiples within a pod.
- **lemon:** A tart yellow citrus fruit with high acidity levels.
- **lentil:** A very small, dried pulse that has been split in half.
- **lime:** A small citrus fruit that can range in color from dark green to yellow-green.
- **loganberry:** A red-purple hybrid berry made by crossbreeding a raspberry and a blackberry.
- **long-grain rice:** A rice that is long and slender and remains light and fluffy after cooking.
- **looseleaf lettuce:** A mild flavored, rich-colored lettuce with a cascade of leaves held loosely together at the root.
- **lotus root:** The underwater root vegetable of an Asian water lily that looks like a solid-link chain about 3 inches in diameter and up to 4 feet in length.
- **lychee:** An exotic drupe covered with a thin, red, inedible shell and has a light-pink to white flesh that is refreshing, juicy, and sweet.
- **mâche:** A small, leafy green with a velvety texture. Also called lamb's lettuce.
- **mandarin:** A small, intensely sweet citrus fruit that is closely related to the orange, but is more fragrant.
- **mandoline:** A special cutting tool with adjustable steel blades used to cut food into consistently thin slices.
- **mango:** An oval or kidney-shaped tropical drupe with orange to orange-yellow flesh.
- **mangosteen:** A round, sweet and juicy exotic fruit about the size of an orange with a hard, thick, dark-purple rind that is inedible.
- **mealy potato:** A type of potato that is higher in starch and lower in moisture than other types of potatoes.
- **medium-grain rice:** A rice that contains slightly less starch than short-grain rice but is still glossy and slightly sticky when cooked.
- **melon:** A type of fruit that has a hard outer rind (skin) and a soft inner flesh that contains many seeds.
- **mesclun greens:** A mix of young greens that range in color, texture, and flavor.
- **Meyer lemon:** A cross between a lemon and an orange.

- **microgreens:** The first sprouting leaves of an edible plant.
- **milled grain:** A refined grain that has been ground into a fine meal or powder.
- **millet:** A small, round, butter-colored grain that is gluten-free.
- **mincing:** Finely chopping an item to yield a very small cut, yet not entirely uniform, product.
- **morel mushroom:** An uncultivated mushroom with a cone-shaped cap that ranges in height from 2–4 inches and in color from tan to very dark brown.
- **mushroom:** The fleshy, spore-bearing body of an edible fungus that grows above the ground.
- **muskmelon:** A round, orange-fleshed melon with a beige or brown, netted rind.
- **mustard green:** A large, dark-green leaf vegetable from the mustard plant that has a strong peppery flavor.
- **Napa cabbage:** An elongated head of crinkly and overlapping edible leaves that are a pale yellow-green color with a white vein. Also known as celery cabbage.
- **nectarine:** A sweet, slightly tart, orange to yellow drupe with a firm, yellow flesh and a large oval pit.
- **new potato:** Any variety of potato that is harvested before the sugar is converted to starch. Also known as an early crop potato.
- **nopal:** The green, edible leaf of the prickly pear cactus.
- **nori:** A thin, purple-black sea vegetable that turns green when it is toasted.
- **oat groat:** An oat grain that only has the husk removed.
- **oats:** Grains derived from the berries of oat grass.
- **oblique cut:** A slicing cut that produces wedge-shaped pieces with two angled sides. Also known as a rolled cut.
- **oca:** A small, knobby tuber that has a potato-like flesh and ranges in flavor from very sweet to slightly acidic. Also known as a New Zealand yam.
- **okra:** A green fruit-vegetable pod that contains small, round, white seeds and a gelatinous liquid.
- **olive:** A small, green or black drupe that is grown for both the fruit and its oil.
- **onion:** A bulb vegetable made up of many concentric layers of fleshy leaves.
- **orange:** A round, orange-colored citrus fruit.
- **oyster knife:** A small knife with a short, dull-edged blade with a tapered point that is used to open oysters.
- **oyster mushroom:** A broad, fanlike or oyster-shaped mushroom that varies in color from white to gray or tan to dark brown.
- **papaya:** A pear- or cylinder-shaped tropical fruit weighing 1–2 pounds with flesh that ranges in color from orange to red-yellow.

KEY TERMS (continued)

- **paring knife:** A short knife with a stiff 2–4 inch blade used to trim and peel fruits and vegetables.
- **parisienne scoop:** A special cutting tool that has a half-ball cup with a blade edge attached to a handle and that is used to cut fruits and vegetables into uniform spheres.
- **parsnip:** An off-white root vegetable 5–10 inches in length.
- **partial tang:** A shorter tail of a knife blade that has fewer rivets than a full tang.
- **passion fruit:** A small, oval-shaped exotic fruit that typically weighs 2–3 ounces and has firm, inedible skin that can be either yellow or purple.
- **pasta:** A term for rolled or extruded products made from a dough composed of flour, water, salt, oil, and sometimes eggs.
- **pattypan:** A round, shallow summer squash with scalloped edges that is harvested when it is no larger than 2–3 inches in diameter.
- **pea:** The edible seed of various plants in the legume family.
- **peach:** A sweet, orange to yellow drupe with downy skin.
- **pear:** A bell-shaped pome with a thin peel and sweet flesh.
- **pearled grain:** A refined grain with a pearl-like appearance that results from having been scrubbed and tumbled to remove the bran.
- **pectin:** A chemical present in all fruits that acts as a thickening agent when it is cooked in the presence of sugar and an acid.
- **peel:** The thick outer rind of a citrus fruit.
- **peeler:** A special cutting tool with a swiveling, double-edged blade that is attached to a handle and is used to remove the skin or peel from fruits and vegetables.
- **Persian melon:** A green muskmelon with finely textured net on the rind. Also known as a patelquat.
- **persimmon:** A bright-orange tropical fruit that grows on trees and is similar in shape to a tomato.
- **pineapple:** A sweet, acidic tropical fruit with a prickly, pinecone-like exterior and juicy, yellow flesh.
- **pith:** The white layer just beneath the peel of a citrus fruit.
- **plantain:** A tropical fruit that is a close relative of the banana, but is larger and has a dark brown skin when ripening.
- **plum:** An oval-shaped drupe that grows on trees in warm climates and comes in a variety of colors such as blue-purple, red, yellow, or green.
- **pome:** A fleshy fruit that contains a core of seeds and has an edible skin.
- **pomegranate:** A round, bright-red tropical fruit with a hard, thick outer skin.

- **porcini mushroom:** An uncultivated, pale-brown mushroom with a smooth, meaty texture and a pungent flavor. Also known as a cèpe.
- **portobello mushroom:** A very large and mature, brown cremini mushroom that has a flat cap measuring up to 6 inches in diameter.
- **potato:** A round, oval, or elongated tuber that is the only edible part of the potato plant.
- **prickly pear:** A pear-shaped tropical fruit with protruding prickly fibers that is a member of the cactus family.
- **pulse:** The dried seed of a legume.
- **pumpkin:** A round fruit-vegetable with a hard orange skin and a firm flesh that surrounds a cavity filled with seeds.
- **purple potato:** A mealy potato with smooth, thin, bluishpurple skin and purple flesh. Also known as a blue potato.
- **quince:** A hard yellow pome that grows in warm climates.
- **quinoa:** A small, round, gluten-free grain that is classified as a complete protein.
- **radicchio:** A small, compact head of red leaves, similar to a small head of red cabbage.
- **radish:** A root vegetable with a white flesh and a peppery taste that comes in many colors and shapes.
- **rambutan:** Fragrant and sweet exotic fruit covered on the outside with soft, hair-like spikes.
- **ramp:** A wild leek with a flavor similar to scallions, yet with more zing.
- **raspberry:** A slightly tart, red aggregate fruit that grows in clusters.
- **rat-tail tang:** A narrow rod of metal that runs the length of a knife handle but is not as wide as the handle.
- **red flame grape:** A seedless grape that ranges from a light purple-red color to a dark-purple color.
- **red potato:** A round, waxy, red-skinned potato with white flesh.
- **refined grain:** A grain that has been processed to remove the germ, bran, or both.
- **rhubarb:** A tart stem vegetable that ranges in color from pink to red and is usually prepared like a fruit.
- **ribbon pasta:** A thin, round strand or flat, ribbonlike strand of pasta.
- **rivet:** A metal fastener used to attach the tang of a knife to the handle.
- **rolled oats:** Oats that have been steamed and flattened into small flakes. Also known as old-fashioned oats.
- **romaine lettuce:** A lettuce with long, green leaves that grow in a loosely packed, elongated head on crisp center ribs.

KEY TERMS (continued)

- **rondelle cut:** A slicing cut that produces disks. Also known as a round cut.
- **root vegetable:** An earthy-flavored vegetable that grows underground and has leaves that extend above ground.
- **russet potato:** A mealy potato with thin brown skin, an elongated shape, and shallow eyes.
- **rutabaga:** A round root vegetable with a yellow-tinted flesh that is the result of a cross between a Savoy cabbage and a turnip.
- **rye:** A hearty grain with dark-brown kernels that are longer and thinner than wheat.
- **salsify:** A white or black root vegetable that can grow up to 12 inches in length.
- **Santa Claus melon:** A large, mottled yellow and green variety of muskmelon that has a slightly waxy skin and soft stem end when ripe.
- **santoku knife:** A knife with a razor-sharp edge and a heel that is perpendicular to the spine.
- **satsuma:** A small, seedless variety of mandarin.
- **Savoy cabbage:** A conical-shaped head of tender, crinkly, edible leaves that are blue-green on the exterior and pale green on the interior.
- **scallion:** A small bulb vegetable with a slightly swollen base and long, slender, green leaves that are hollow. Also known as a green onion.
- **scimitar:** A long knife with an upward curved tip that is used to cut steaks and primal cuts of meat.
- **sea vegetables:** Edible saltwater plants that contain high amounts of dietary fiber, vitamins, and minerals.
- **Seckel pear:** A small pear that is sometimes called a honey pear or sugar pear because of its syrupy, fine-grained flesh and complex sweetness.
- **semolina:** The granular product that results from milling the endosperm of durum wheat.
- **shallot:** A very small bulb vegetable that is similar in shape to garlic and has two or three cloves inside.
- **shaped pasta:** A pasta that has been extruded into a complex shape such as a corkscrew, bowtie, shell, flower, or star.
- **shiitake mushroom:** An amber, tan, brown, or dark-brown mushroom with an umbrella shape and curled edges. Also known as a forest mushroom.
- **short-grain rice:** Rice that is almost round in shape and has moist grains that stick together when cooked.
- **slicer:** A knife with a narrow blade 10–14 inches long that is used to slice roasted meats. Also known as a carving knife.
- **sorrel:** A large, green, leaf vegetable that ranges in color from pale green to dark green and from 2–12 inches in length.
- **spaghetti squash:** A dark-yellow winter squash with pale-yellow flesh that can be separated into spaghetti-like strands after it is cooked.
- **spelt:** An ancient grain with a nutty flavor and high protein content that is also a good source of riboflavin, zinc, and dietary fiber.
- **spinach:** A dark-green leaf vegetable with a slightly bitter flavor that may have flat or curly leaves, depending on the variety.
- **spine:** The unsharpened top part of the knife blade that is opposite the edge.
- **squash blossom:** The edible flower of a summer or a winter squash.
- **star fruit:** An exotic fruit that is shaped like a star when cut. Also known as a carambola.
- **steel:** A steel rod approximately 18 inches long attached to a handle and used to align the edge of knife blades. Also known as a butcher's steel.
- **steel-cut oats:** Oat groats that have been toasted and cut into small pieces.
- **stem vegetable:** The main trunk of a plant that develops buds and shoots instead of roots.
- **strawberry:** A bright-red, heart-shaped berry covered with tiny black seeds.
- **stuffed pasta:** A pasta that has been formed by hand or machine to hold fillings.
- **summer squash:** A fruit-vegetable that grows on a vine and has edible skin, flesh, and seeds.
- **sunchoke:** A tuber with thin, brown, knobby-looking skin. Also known as a Jerusalem artichoke.
- **sweet corn:** A fruit-vegetable that has edible seeds called kernels that grow in rows on a spongy cob encased by thin leaves (husks), forming what is referred to as an ear of corn.
- **sweet potato:** A tuber that has a paper-thin skin and flesh that ranges in color from ivory to dark orange.
- **tang:** The unsharpened tail of a knife blade that extends into the handle.
- **tangelo:** A hybrid of a tangerine and a grapefruit.
- **tangerine:** A small citrus fruit with a slightly red-orange peel that can be easily removed without a knife.
- **tangor:** A hybrid of a tangerine and a sweet orange.
- **tatsoi:** A spoon-shaped, emerald-colored leaf vegetable native to Japan.
- **Thompson grape:** A seedless grape that is pale to light green in color.
- **tip:** The front quarter of a knife blade.
- **tomatillo:** A small fruit-vegetable with a thin, papery husk covering a pale-green skin that encases a pale-green flesh. Also known as a Mexican husk tomato.

KEY TERMS (continued)

- **tomato:** A juicy fruit-vegetable that contains edible seeds.
- **tourné cut:** A carved, football-shaped cut with seven sides and flat ends.
- **tourné knife:** A short knife with a curved blade that is primarily used to carve vegetables into a specific shape called a tourné, which is a seven-sided football shape with flat ends. Also known as a bird's beak knife.
- **tropical fruit:** A type of fruit that comes from a hot, humid location but is readily available.
- **tube pasta:** A pasta that has been pushed through an extruder and then fed through a cutter that cuts the tubes to desired length.
- **tuber:** A short, fleshy vegetable that grows underground and bears buds capable of producing new plants.
- **turnip:** A round, fleshy root vegetable that is purple and white in color.
- **turnip green:** A dark-green leaf vegetable that grows out of the top of the turnip root vegetable.
- **ugli fruit:** A large, teardrop-shaped, seedless citrus fruit. Also known as a uniq fruit.
- **utility knife:** A multipurpose knife with a stiff 6–10 inch blade that is similar in shape to a chef's knife but much narrower at the heel.
- **variety fruit:** A fruit that is the result of breeding two or more fruits of the same species that have different characteristics.
- **vegetable:** An edible root, bulb, tuber, stem, leaf, flower, or seed of a nonwoody plant.
- **wakame:** A long, tender, grayish-green sea vegetable that expands to seven times its size when soaked in water.
- **water chestnut:** A small tuber with brownish-black skin and white flesh. Also known as a water caltrop.
- **watercress:** A small, crisp, dark-green, leaf vegetable with a pungent, yet slightly peppery flavor.

- **watermelon:** A sweet, extremely juicy melon that is round or oblong in shape, with pink, red, or golden flesh and green skin.
- **waxy potato:** A type of potato with a thin skin and slightly waxy flesh that is lower in starch and higher in moisture than mealy potatoes.
- **wheat:** A light yellow cereal grain cultivated from an annual grass that yields the flour used in many pastas and baked goods.
- **whetstone:** A stone used to grind the edge of a blade to the proper angle for sharpness.
- **white potato:** An oblong mealy potato with a thin, white or light-brown skin and tender, white flesh.
- **whole grain:** A grain that only has the husk removed.
- **winter squash:** A fruit-vegetable that grows on a vine and has a thick, hard, inedible skin and firm flesh surrounding a cavity filled with seeds.
- **wood ear mushroom:** A brownish-black, ear-shaped mushroom that has a slightly crunchy texture. Also known as a cloud ear or a tree ear mushroom.
- **yam:** A large tuber that has thick, barklike skin and flesh that varies in color from ivory to purple.
- **yellow potato:** An oval, waxy potato with thin, yellowish skin and flesh and pink eyes.
- **yellow squash:** A summer squash that resembles a bowling pin with either a straight or crooked neck.
- **zest:** The colored, outermost layer of the peel of a citrus fruit that contains a high concentration of oil.
- **zester:** A special cutting tool with tiny blades inside of five or six sharpened holes that are attached to a handle.
- **zucchini:** An elongated summer squash that resembles a cucumber and is available in green or yellow varieties.

Refer to CD-ROM for **Quick Quiz®** questions

Sustainability Corner

The restaurant business is one of the largest consumers of energy, with most of that energy being used by cooking equipment, food storage equipment, and dishwashers. The average professional kitchen uses five times more energy than the rest of the building. The cost of energy is a factor that affects the budget and profit margins of virtually all foodservice operations. Fortunately, there are a number of ways to provide high-quality food and service while still being energy efficient, conserving natural resources, and improving profits.

One simple yet effective strategy for reducing energy costs is to develop a power up/power down schedule. For example, broilers do not have a thermostatic control and can run at full power all day. However, according to the Food Service Technology Center, broilers operating 12 hours per day without an effective power up/power down schedule can add an extra $7344 per year to energy costs.

Also, lighting in a professional kitchen can be extensive, expensive, and heat producing, especially in an already hot kitchen. Traditional light bulbs use only 10% of their energy to produce light, with the remaining 90% burnt off as heat. Replacing traditional incandescent light bulbs with compact fluorescent light bulbs, or CFLs, is one of the easiest ways to shrink a power bill and stay cooler in the kitchen. Replacing 24-hour EXIT signs with CFLs and equipping the signs with off/on timers is one small change that can make a substantial difference.

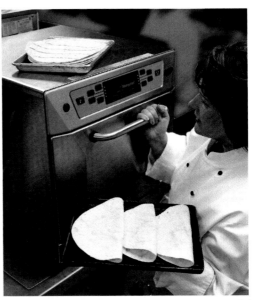

Merrychef

A major strategy for saving energy and reducing costs is to invest in energy-efficient equipment and appliances. Using energy-efficient ovens, fryers, griddles, hot food holding cabinets, refrigerators, freezers, ice machines, and dishwashers in the professional kitchen can result in substantial savings. For example, an energy-rated commercial dishwasher can potentially reduce energy bills by $900 per year and water bills by $200 per year.

Another piece of kitchen equipment that is a huge consumer of energy is the ice machine. If a commercial ice machine has earned the ENERGY STAR™ rating, savings can average 15% on energy and 10% on the amount of water used. Installing a timer and shifting ice production to nighttime off-peak hours will make the ice machine even more efficient.

The type of ovens used can also significantly impact energy use. Convection ovens conserve more energy than conventional ovens because they cook food faster at lower temperatures. Minimizing oven preheating times will also conserve energy. Fifteen minutes is generally enough time to preheat an oven. Efficient use of ovens also includes scheduling cooking times so the oven is fully loaded, keeping the oven door closed as much as possible, and turning ovens off when not in use.

Energy Efficient Kitchens

1. ___ vegetables include beets, carrots, celeriac, jicamas, lotus roots, parsnips, radishes, rutabagas, salsify, and turnips.
   A. Leaf
   B. Stem
   C. Root
   D. Bulb

2. The USDA voluntary grading system for ___ vegetables includes U.S. Extra Fancy, U.S. Fancy, U.S. Extra No. 1, U.S. No. 1, U.S. No. 2, and U.S. No. 3.
   A. fresh
   B. canned
   C. frozen
   D. dried

3. The ___ is the unsharpened top part of a knife blade that is opposite the edge.
   A. tip
   B. heel
   C. spine
   D. bolster

4. A ___, also known as a stone fruit, is a type of fruit that contains one hard seed or pit.
   A. melon
   B. berry
   C. drupe
   D. pome

5. The USDA has a voluntary grading program for fresh ___ based on uniformity of shape, size, color, texture, and the absence of defects.
   A. potatoes
   B. fruit
   C. grains
   D. vegetables

6. Adding ___ or an acid such as lemon juice to fruit can help prevent it from becoming mushy in the cooking process.
   A. salt
   B. pepper
   C. flour
   D. sugar

7. The conventional clock system of placing vegetables at ___ o'clock, proteins at 6 o'clock, and starches at 11 o'clock is a good method to follow when plating.
   A. 2
   B. 4
   C. 8
   D. 12

8. When properly positioning a knife, the side of the blade should rest against the knuckle of the ___ finger of the guiding hand.
   A. index
   B. middle
   C. ring
   D. pinkie

9. A ___ cut is a stick cut that produces a stick-shaped item ⅛ × ⅛ × 2 inches long.
   A. julienne
   B. fine julienne
   C. batonnet
   D. fine batonnet

10. When making ___, thin pasta dough is cut into circles and a filling is placed in the center of each circle.
    A. ravioli
    B. tortellini
    C. capellini
    D. manicotti

11. ___ potatoes are the preferred type of potato for baking, frying, mashing, puréeing, and making casseroles.
    A. Mealy
    B. Waxy
    C. Red
    D. Purple

12. Fresh potatoes must be kept in a ___ place that allows them to breathe.
    A. dry, warm, dark
    B. moist, warm, dark
    C. dry, cool, dark
    D. moist, cool, dark

13. ___ is a small, round, gluten-free grain that is classified as a complete protein.
    A. Barley
    B. Buckwheat
    C. Quinoa
    D. Spelt

14. The ___ is the largest component of a grain kernel and consists of carbohydrates and a small amount of protein.
    A. husk
    B. bran
    C. germ
    D. endosperm

15. ___ rice is rice that has had only the husk removed.
    A. White
    B. Brown
    C. Short-grain
    D. Long-grain

16. Cutting open ___ potatoes is the only way to determine if they are done.
    A. baked
    B. fried
    C. grilled
    D. simmered

17. ___ is the most common method of cooking grains.
    A. Simmering
    B. Boiling
    C. Sautéing
    D. Stir-frying

18. Cellentani, ditalini, elbows, manicotti, penne, pipettes, rigatoni, and ziti are common types of ___ pasta.
    A. shaped
    B. ribbon
    C. tube
    D. stuffed

19. A tourné cut is a carved, football-shaped cut with ___ sides and flat ends.
    A. five
    B. six
    C. seven
    D. eight

20. When pasta is cooked al dente, there should be ___ resistance in the center of the pasta when it is chewed.
    A. no
    B. a slight
    C. a lot of
    D. complete

21. Describe the procedure for blanching vegetables.

    _____

    _____

    _____

    _____

    _____

22. Describe proper knife grip and positioning.

    _____

    _____

    _____

    _____

    _____

23. Describe the procedure for preparing risotto.

    _____

    _____

    _____

    _____

    _____

24. Describe the procedure for preparing tortellini.

    _____

    _____

    _____

    _____

    _____

    _____

    _____

SKETCHING EXERCISE

25. Sketch a chef's knife and label each part.

Notes